"十三五"江苏省高等学校重点教材(编号:2016-2-030)
全国高等职业教育"十三五"规划教材

Android 应用开发项目实战

主　编　刘贤锋　孙华林
副主编　鲍建成　盛昀瑶
参　编　陈爱民　廖定安　顾　飞
主　审　王继水

机械工业出版社

本书涵盖了 Android 开发的基础概念、技术知识、实践应用的每一个领域。从初学者的角度，以丰富的实例、案例，通俗易懂的语言，简单的图示，系统全面地讲述了 Android 开发中所用的技术。书中的所有案例都是基于 Android 4.2 版本并在 Eclipse（ADT）集成开发环境下成功调试的，读者可以将书中的所有案例源码导入到 Eclipse 中运行及调试。本书详细介绍了 Android 应用开发的相关基础知识，内容覆盖了 Android 开发环境搭建与配置、开始我的第一个 Android 应用程序、基础 UI 组件在 APP 界面中的运用、各种 Android 资源在 APP 应用中的引用、使用 Intent 实现界面跳转、高级 UI 组件的应用、Android 数据存储与 SQLite 数据库应用、最后给出了两个具体的 APP 实践应用——欧洲杯信息采集 APP 和新闻客户端 APP 应用。

本书不仅可以作为各级各类学校、培训机构的移动互联专业的教材，也可作为 Android 应用程序开发人员入门学习时的参考资料。

本书提供配套的电子课件和程序代码源文件，需要的教师可登录 www.cmpedu.com 进行免费注册，审核通过后即可下载；或者联系编辑索取（QQ：1239258369，电话：010-88379739）。

图书在版编目（CIP）数据

Android 应用开发项目实战/刘贤锋，孙华林主编.
--北京：机械工业出版社，2017.8
全国高等职业教育"十三五"规划教材
ISBN 978-7-111-57736-2

Ⅰ. ①A… Ⅱ. ①刘… ②孙… Ⅲ. ①移动终端-应用程序-程序设计-高等职业教育-教材 Ⅳ. ①TN929.53

中国版本图书馆 CIP 数据核字（2017）第 192308 号

机械工业出版社（北京市百万庄大街 22 号　邮政编码 100037）
策划编辑：李文轶　　责任编辑：李文轶
责任校对：张艳霞　　责任印制：孙　炜
北京中兴印刷有限公司印刷
2017 年 9 月第 1 版・第 1 次印刷
184mm×260mm・17.25 印张・418 千字
0001–3000 册
标准书号：ISBN 978-7-111-57736-2
定价：46.00 元

凡购本书，如有缺页、倒页、脱页，由本社发行部调换

电话服务　　　　　　　　　　　　网络服务
服务咨询热线：(010)88379833　　机　工　官　网：www.cmpbook.com
　　　　　　　　　　　　　　　　机　工　官　博：weibo.com/cmp1952
读者购书热线：(010)88379649　　教育服务网：www.cmpedu.com
封面无防伪标均为盗版　　　　金　书　网：www.golden-book.com

前　言

在以开放手机联盟（Open Handset Alliance，OHA）和 Google 公司为先驱的开拓下，Android 作为一个热门、新潮、免费、开放的移动互联网平台在业界刮起了一阵"旋风"：Android 系统已经成为一个开放式的手机和平板电脑的操作系统，市场占有率也在稳步上升。而这本书可以为初学移动应用开发的读者提供设计、开发、测试、调试、发布 Android APP 应用程序等一系列的指导。

一、本书的结构

本书共分为两个部分，即基础篇和应用篇，基础篇共 7 个项目，应用篇有两个项目。

项目一　Android 开发环境搭建及配置：本模块在给出 Java 开发环境及 Android 可视化开发平台 ADT 的搭建基础上，重点介绍了使用真机和 AVD 作为 Android APP 运行及调试的方法。

项目二　开始我的第一个 Android 应用程序：本项目主要介绍在 ADT 平台下开发并运行第一个 Android 应用程序的方法和步骤，并在此基础上重点介绍了使用 DDMS 对应用程序进行调试的步骤和注意事项，分析了 Android 应用程序的结构。最后给出了在 Eclipse 中对 Android 应用进行签名及使用 AAPT 打包资源的方法。

项目三　基础 UI 组件在 APP 界面中的运用：本项目主要在介绍 View 类的基础上讲解了 Android 界面编程的 3 种模式，即使用 XML 布局文件搭建 UI 界面、使用 Java 代码实现 UI 界面及使用 XML 和 Java 代码混合实现 UI 界面的方法，并给出了自定义一个 View 的步骤。接着介绍了 Android 常用的基础 UI 组件的使用技巧和方法，如界面的布局方式、按钮、文本框等一些简单组件的使用。最后结合案例，尤其是用户注册界面的实现，重点讲述使用网格布局管理器对界面进行布局的方法和步骤。

项目四　各种 Android 资源在 APP 应用中的引用：本项目详细讲解了 Android 中的各种资源及使用。资源是 Android 应用中的重要组成部分。Android 应用程序可以通过使用各种资源来设置颜色、字体大小、风格等属性，也可以通过资源存储图像、数组等常用资源。

项目五　使用 Intent 实现界面跳转：Android 应用程序主要由 4 个部分组成，分别为活动（Activity）组件、服务（Service）组件、内容提供者（Content Provider）和广播接收者组件（Broadcast Receiver）。其中，Activity 组件是 Android 应用的入口，是 Android 初级程序员必须掌握的组件之一。本部分重点讲解了 Activity 组件的概念、建立、配置及使用方法，在此基础上还给出了使用 Bundle 在不同 Activity 之间进行数据交换的方法，并讨论了 Activity 的生命周期，最后给出了 Android 编程的事件处理机制。

项目六　高级 UI 组件的应用：本项目主要介绍了 AdapterView 及其子类 ListView、AutoCompleteTextView、GridView、ExpandableListView 等列表的用法。在介绍上述组件的同时，也重点介绍了 Adapter 适配器及其子类 ArrayAdapter、SimpleAdapter 等用法。最后还详细介绍了其他一些常用组件（如 ProgressBar、DatePickerDialog、SearchView、TabHost 及 AlertDialog 等）的用法。

项目七　Android 数据存储与 SQLite 数据库应用：本项目首先主要介绍了应用程序首选项 SharedPreferences 及如何读取首选项中的数据；接下来介绍了文件存储、Android 内部存储数据及外部存储数据的读/写方法；最后重点讲述了 SQLite 数据库，介绍了使用 SQLiteDatabase 类和 SQLiteOpenHelper 类操作数据的步骤和方法，并讲解了 SQLite3 工具的用法。

项目八　欧洲杯信息采集 APP 应用：本项目基于 Android 平台对欧洲杯信息采集系统进行研发与制作，将整个项目拆分成单独模块，系统性实践 Android 界面编程在实际业务当中的运用。

项目九　新闻客户端 APP 应用：本项目重点介绍了 Android 网络编程技术。它结合服务器端新闻发布系统，基于 HttpURLConnection 访问服务器，采用自定义数据适配器、异步加载网络图片、JSON 轻量级数据解析技术，构建了 Android 移动新闻客户端 APP 应用。

二、本书的特点

（1）内容实用，适用性强；
（2）讲解详细，容易上手；
（3）全书基于"任务驱动"的讲解模式；
（4）以提高动手能力为核心。

三、本书使用的 Android 开发环境

本书的所有 Android 应用程序代码都基于如下环境编写：
（1）Windows 7 的 32 位操作系统；
（2）Eclipse ADT v21.0.0；
（3）Java SE 开发工具包 JDK 1.6；
（4）Android SDK Version 4.2，API Level 17；
（5）Android 手机设备，例如 HTC Nexus One、红米、小米 3S、三星 S5。

四、本书适合的读者

本书涵盖了 Android 基础开发所涉及的概念、技术知识点、案例实践及一些开发经验。本书主要针对以下两类人群。

（1）高校、职业技术院校、培训机构的讲师和学生

当前，Android 移动互联应用的"热门"使得大中专院校、培训机构等都开设了相近的专业和培训班，本书针对教师的"教"与学生的"学"的特点组织内容，更加适合学生和老师。

（2）想要学习 Android 应用程序开发的初学者

本书主要针对有 Java 程序开发经验而没有 Android 移动开发经验的人员，这本书中的丰富、实用的案例将会带领读者快速进入移动开发领域。

书中不足和疏漏之处还请读者批评指正。

编　者

目 录

前言

第一篇 基 础 篇

项目一 Android 开发环境搭建及配置 …… *1*
 模块一 搭建 Android 开发环境 ……… *1*
  任务 1 搭建 Java 开发环境 …………… *2*
  任务 2 搭建 Android 开发平台 IDE …… *8*
 模块二 安装、运行及调试环境……… *12*
  任务 1 使用真机作为运行及调试环境的参数配置 …………………… *12*
  任务 2 使用 AVD 作为运行及调试环境的参数配置 …………………… *13*
项目二 开始我的第一个 Android 应用程序……………………………… *16*
 模块一 创建 Android 应用工程 ……… *16*
  任务 1 基于 ADT 新建 Android 项目 …… *17*
  任务 2 Android 应用程序结构分析……… *20*
  任务 3 资源文件的使用方式 …………… *23*
 模块二 运行及调试 APP 应用 ……… *24*
  任务 1 启动 AVD …………………… *25*
  任务 2 运行 APP …………………… *25*
  任务 3 使用 DDMS 进行调试 ………… *27*
 模块三 签名并打包 Android 应用程序……………………………… *28*
  任务 APP 签名及打包 ……………… *29*
项目三 基础 UI 组件在 APP 界面中的运用 …………………… *37*
 模块一 使用 XML 和 Java 代码混合实现 UI 界面 ………………… *37*
  任务 1 使用 XML 布局文件搭建 UI 界面 ………………………… *38*
  任务 2 使用 Java 代码实现 UI 界面 …… *41*

  任务 3 使用 XML 和 Java 代码实现图片浏览器 …………………… *43*
 模块二 继承 View 类实现自定义 View 组件 …………………… *46*
  任务 自定义 View 组件实现可以随意拖动的小球 ………………… *46*
 模块三 使用 UI 布局管理器实现界面布局 ………………………… *51*
  任务 1 使用线性布局实现在界面中动态添加按钮 ………………… *52*
  任务 2 使用表格布局实现窗口布局 …… *56*
  任务 3 使用网格布局实现一个简易的计算器 …………………… *58*
 模块四 使用基础 UI 组件实现用户注册界面 ……………………… *61*
  任务 用户注册界面的具体实现 ……… *62*
项目四 各种 Android 资源在 APP 应用中的引用………………… *76*
 模块一 字符串/颜色/数组等基础资源的定义和使用 ………… *76*
  任务 1 基于资源引用方式优化用户注册页面 …………………… *77*
  任务 2 数组资源的使用 …………… *89*
  任务 3 Drawable 资源的使用 ………… *93*
 模块二 Android 系统资源及 assets 资源的使用………………… *96*
  任务 1 Android 系统资源的访问和使用 ………………… *96*
  任务 2 assets 资源的使用 ………… *97*

V

项目五　使用 Intent 实现界面跳转 …… 99
　模块一　Activity 组件的创建/
　　　　　启动/配置 …………………… 99
　　任务 1　Activity 组件的创建及配置 …… 100
　　任务 2　使用 Intent 启动 Activity 实现
　　　　　界面跳转 ………………………… 106
　　任务 3　Activity 组件生命周期的验证 …… 111
　模块二　使用 Bundle 实现界面间
　　　　　参数传递 ………………………… 116
　　任务　数据传递的具体实现 ……………… 116
　模块三　Android 事件处理编程 ………… 121
　　任务 1　内部类实现事件监听器 ………… 122
　　任务 2　匿名内部类实现事件
　　　　　监听器 …………………………… 123
　　任务 3　外部类实现事件监听器 ………… 125
　　任务 4　Activity 实现事件监听器 ……… 126
　　任务 5　为组件绑定相关属性实现
　　　　　事件监听器 ……………………… 127
　　任务 6　Handler 消息传递编程 ………… 127

项目六　高级 UI 组件的应用 ………… 134
　模块一　使用 ListView 显示列表
　　　　　数据 ……………………………… 134
　　任务 1　直接继承 ListActivity 创建
　　　　　ListView ………………………… 135
　　任务 2　使用 XML 布局文件创建
　　　　　ListView ………………………… 139
　　任务 3　使用 ArrayAdapter 创建
　　　　　ListView ………………………… 140
　　任务 4　使用 SimpleAdapter 创建
　　　　　ListView ………………………… 141

　模块二　文本框输入中自动提示
　　　　　列表的实现 ……………………… 146
　　任务　基于 AutoCompleteTextView 实现自动
　　　　　提示列表 ………………………… 146
　模块三　图片浏览器的实现 ……………… 156
　　任务 1　用 GridView 实现带预览功能的
　　　　　图片浏览器 ……………………… 156
　　任务 2　用 AdapterViewFlipper 实现自动
　　　　　播放图片 ………………………… 160
　模块四　对话框/日期选择框等常用
　　　　　对话框的创建 …………………… 163
　　任务 1　使用 AlertDialog 实现对话框 …… 164
　　任务 2　使用 DatePickerDialog 实现
　　　　　日期输入 ………………………… 167
　　任务 3　使用 TabHost 选项卡模拟手机
　　　　　通话记录界面 …………………… 171

**项目七　Android 数据存储与 SQLite
　　　　　数据库应用** ……………………… 176
　模块一　Android 数据存储操作 ………… 176
　　任务 1　使用 SharedPreferences 设置
　　　　　系统参数 ………………………… 177
　　任务 2　Android 内存数据读/写操作 …… 181
　　任务 3　Android SD 卡数据读/写操作 …… 185
　模块二　SQLite 数据库操作 ……………… 190
　　任务 1　使用 SQLiteDatabase 实现
　　　　　SQLite 数据库操作 ……………… 190
　　任务 2　使用 SQLiteOpenHelper 实现
　　　　　SQLite 数据库操作 ……………… 196
　　任务 3　使用 SQLite3 工具操作
　　　　　数据库 …………………………… 205

第二篇　应　用　篇

项目八　欧洲杯信息采集 APP 应用 …… 207
　模块一　欧洲杯主视图界面的实现 ……… 207
　　任务 1　构建欧洲杯主界面 ……………… 208
　　任务 2　列表呈现国家名和图标 ………… 212

　模块二　积分榜的实现 …………………… 223
　　任务 1　积分榜对话框的创建及弹出 …… 223
　　任务 2　列表显示小组内国家积分
　　　　　情况 ……………………………… 225

| 模块三　赛事明细列表的实现 ……… *231* | 模块二　新闻栏目列表功能的实现 … *253* |

　　任务1　以列表形式呈现赛事明细 ……… *232*　　　　任务1　准备工作：测试服务器端新闻
　　任务2　以列表形式呈现可折叠的　　　　　　　　　　　　　栏目列表API接口 ……… *254*
　　　　　　积分榜明细 ……………… *234*　　　　任务2　构建新闻栏目列表界面 ……… *255*
项目九　新闻客户端APP应用 ……… *239*　　　　任务3　用JSON轻量级数据解析技术
　模块一　用户登录功能的实现 ……… *239*　　　　　　　　实现新闻栏目列表功能 ……… *256*
　　任务1　准备工作：服务器端应用　　　　　模块三　新闻栏目列表功能的实现 … *260*
　　　　　　程序的部署 ……………… *240*　　　　任务1　准备工作：测试服务器端新闻
　　任务2　编写配置文件及网络　　　　　　　　　　　　　栏目列表API访问接口 ……… *261*
　　　　　　访问工具 ………………… *244*　　　　任务2　构建新闻栏目列表界面 ……… *262*
　　任务3　构建用户登录界面 ………… *246*　　　　任务3　新闻栏目列表功能实现 ……… *262*
　　任务4　基于HttpURLConnection实现　　**参考文献** ……………………………… *268*
　　　　　　APP客户端用户登录 ……… *249*

第一篇 基 础 篇

项目一 Android 开发环境搭建及配置

Android（安卓）系统是手机或一些平板电脑等移动终端的操作系统，可以说是现在非常流行的智能终端系统之一。三星、华为、HTC 等手机厂商早已通过 Android 的应用取得了巨大成功。对 Android 开发系统的人才需求也在迅速增长，从趋势上看，Android 软件人才的需求会越来越大。

本项目将依托功能强大且运行高效、稳定的 Android 4.2 版本，首先简要介绍 Android 平台的发展史、现状，并在此基础上重点介绍开发环境的搭建、运行及调试方法。

【知识目标】

- Android 的发展和历史
- Android 平台架构和特性
- 搭建 Android 开发环境的步骤和方法
- 使用真机作为运行及调试环境的配置步骤

【模块分解】

- 搭建 Android 开发环境
- 安装、运行及调试环境

模块一 搭建 Android 开发环境

【模块描述】

要快速掌握 Android 应用开发，首先就需要正确搭建 Android 开发及运行环境。目前流行的 APP 应用开发使用 IDE 工具较多，其中以谷歌的 Android Studio 和基于开源软件 Eclipse 的 ADT 插件最为流行。本模块将选择基于 Eclipse 的 ADT 插件来搭建 Android 开发平台。

搭建基于 Eclipse 的 ADT 的 Android APP 开发平台过程非常简单，大致有以下几个步骤。

1) 安装及配置 JDK。
2) 搭建 Android 开发 IDE，具体包括：搭建 Eclipse 开发环境；下载及安装对应版本的

Android SDK；为 Eclipse 安装插件。

知识点	技能点
➢ Android 开发平台及开发环境 ➢ Android APP 应用结构 ➢ Android SDK 版本及核心 API ➢ Eclipse 常见插件及功能	➢ JDK 的安装及环境变量的配置 ➢ 插件集成方法及技巧 ➢ 真机调试环境的设置 ➢ AVD 虚拟机的设置及启动

任务1　搭建 Java 开发环境

【任务描述】

JDK（Java Developer's Kit）即 Java 开发工具包，有时也被称为 J2SDK。该软件工具包含 Java 语言的编译工具、运行工具及软件运行环境（JRE）。JDK 是 Sun 公司（目前已经被 Oracle 公司收购）提供的一款免费的以 Java 语言为基础的开发工具，在安装其他开发工具之前，必须首先安装 JDK，本书采用 JDK 1.6 版本。

【任务实施】

（1）JDK 的下载

首先获取 JDK（可以到官方网站上下载）：

http://www.oracle.com/technetwork/java/javase/downloads/index.html

在浏览器中输入上面的地址，进入如图 1-1 所示的界面。

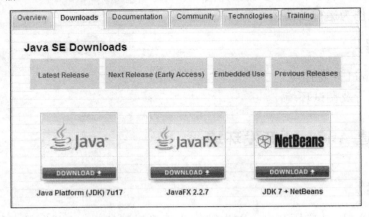

图 1-1　下载 JDK

单击图 1-1 中的 Latest Release 按钮，可以进入如图 1-2 所示的下载页面。

单击图 1-2 中的按钮 ![JDK DOWNLOAD]，进入如图 1-3 所示的下载页面，单击相应内容即可下载。

（2）JDK 的安装步骤

1）下载完毕后，双击图 1-3 中的 jdk-6u43-windows-i586.exe 文件，进入自动解压界面。

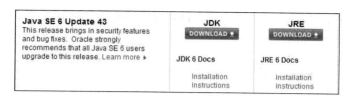

图 1-2　下载 JDK

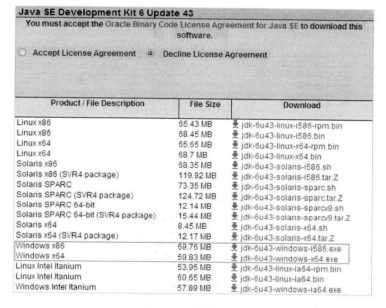

图 1-3　下载 JDK

2）解压完后，进入许可证协议界面，如图 1-4 所示。

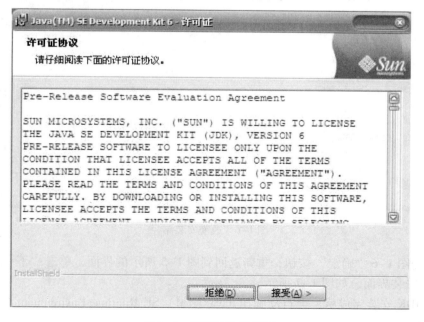

图 1-4　许可证协议界面

3）单击图1-4中"接受"按钮，进入自定义安装界面，选择要安装的组件（默认全部安装）和路径，如图1-5所示。

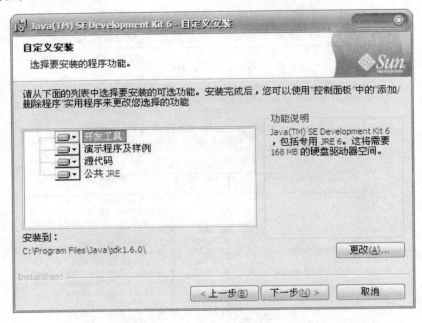

图1-5　自定义安装界面

4）单击图1-5中的"更改"按钮，改变安装路径，如图1-6所示。

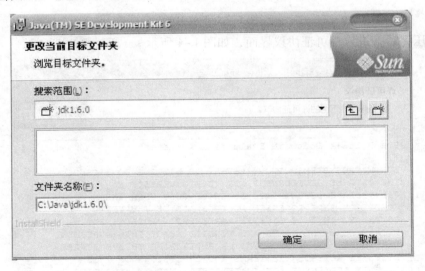

图1-6　改变安装路径

5）单击图1-6"确定"按钮，重新返回到图1-5所示的界面，单击"下一步"按钮即可进入正在安装界面，如图1-7所示。

6）在JDK安装完成后，会自动弹出Java（TM）SE Runtime Environment 6的界面，如图1-8所示。

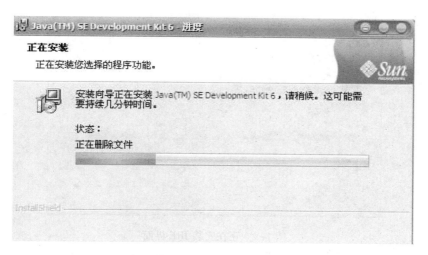

图 1-7　正在安装界面

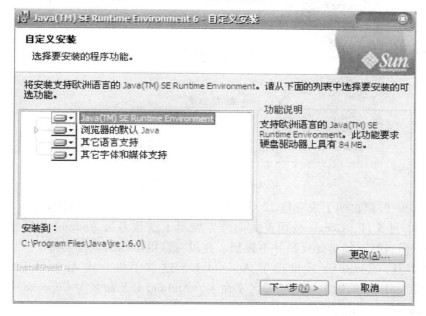

图 1-8　自定义安装 JRE 界面

7）如图 1-8 所示单击"更改"按钮，改变 JRE 的安装路径（本书安装路径为 C:\Program Files\java\jre1.6.0）后单击"下一步"按钮，即可进入正在安装 JRE 界面，如图 1-9 所示。

8）在弹出的安装完成界面中单击"完成"按钮，完成安装，如图 1-10 所示。

【知识学习】Android 概述

（1）Android 的发展与历史

Android 是由 Andy Rubin 创立的一个手机操作系统，后来被 Google（谷歌）公司收购，目前已经被 Google 公司打造成为一个标准化、开放式的移动终端软件平台。

2003 年 10 月，HAndy RubinH 等人创建 Android 公司，并组建 Android 团队。2005 年 8

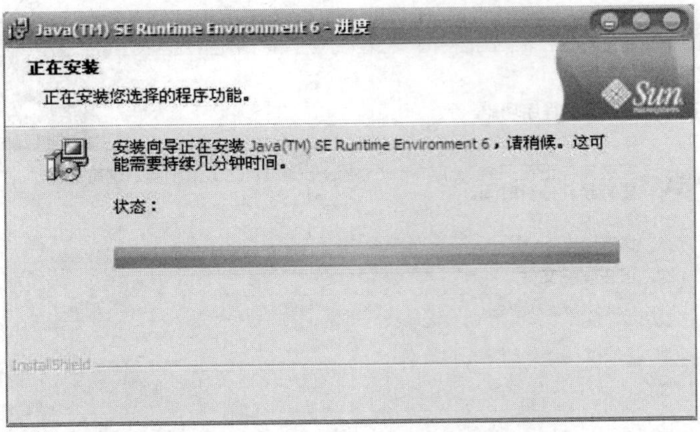

图 1-9　正在安装 JRE 界面

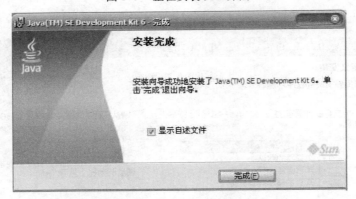

图 1-10　安装完成界面

月 17 日，Google 低调收购了成立仅 22 个月的高科技企业 Android 及其团队。

2007 年 11 月 5 日，Google 公司正式向外界展示了这款名为 Android 的操作系统，发布了 1.0 版本，但当时市场对此反映并不强烈，并没有赢得市场的广泛支持。

2009 年 4 月，Google 正式推出了 Android 1.5 这款手机，从 Android 1.5 版本开始，Google 开始将 Android 的版本以甜品的名字命名，Android 1.5 命名为 Cupcake（纸杯蛋糕）。该系统与 Android 1.0 相比有了很大的改进。

2009 年 9 月份，Google 发布了 Android 1.6 的正式版，并且推出了搭载 Android 1.6 正式版的手机 HTC Hero（G3），凭借着出色的外观设计及全新的 Android 1.6 操作系统，HTC Hero（G3）成为当时全球最受欢迎的手机。Android 1.6 也有一个有趣的甜品名称，它被称为 Donut（甜甜圈）。

经过两年的发展，2011 年 9 月份，Android 系统的应用数量已经达到了 48 万。在智能手机市场，Android 系统的占有率已经达到了 43%，Android 的市场占有率已经远远超过了当时的苹果公司的 iOS 及微软的 Windows Phone 8。接下来，Android 的版本更新得更快，目前市场上的最新版本为 4.4，本书主要基于 4.2 稳定版。

（2）Android 平台

Android 操作系统平台采用了整合的策略思想，包括底层的 Linux 操作系统、中间层的中间件和上层的 Java 应用程序。

1）Android 的平台特性。

① 应用程序框架支持组件的重用与替换。在开发时可以有选择地安装喜欢的应用程序。

② Dalvik 虚拟机专门对移动设备进行了优化。Android 应用程序（将由 Java 编写、编译的类文件）通过 DX 工具被转换成一种扩展名为 .dex 的文件来执行。Dalvik 虚拟机是基于寄存器的，相对于 Java 虚拟机速度要快很多。

③ 内部集成浏览器基于开源的 WebKit 引擎。有了内置的浏览器，这意味着 WAP 应用的时代即将结束，真正的移动互联网时代已经来临，手机就是一台"小电脑"，可以在网上随意遨游。

④ 优化的图形库包括 2D 和 3D 图形库，3D 图形库基于 OpenGL ES 1.0。强大的图形库给游戏开发带来福音。

⑤ SQLite 被用作结构化的数据存储。

⑥ 提供多媒体支持，包括常见的音频、视频和静态印象文件格式。

⑦ 丰富的开发环境包括设备模拟器、调试工具、内存及性能分析图表和 Eclipse 集成的开发环境插件。

Google 提供的 Android 开发包 SDK 包含了大量的类库和开发工具，并且针对 Eclipse 的可视化开发插件 ADT。

2）Android 平台架构。

图 1-11 所示为 Android 系统的体系架构。

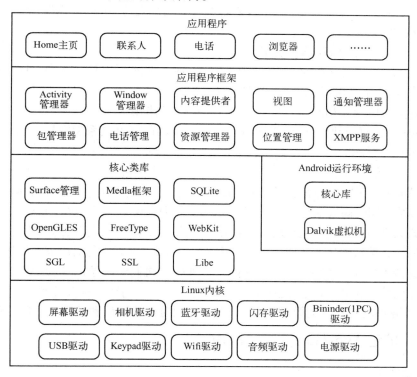

图 1-11　Android 操作系统的体系结构

从上图可以看出，Android 操作系统的体系结构可分为 4 层，由上到下依次是应用程序、应用程序框架、核心类库（Android 程序库和 Android 运行库）和 Linux 内核。其中，第三层还包括 Android 运行时的环境。下面分别来讲解各个部分。

① 应用程序。Android 连同一个核心应用程序包通常会一起被发布，该应用程序包包括 E-mail 客户端、SMS 短消息程序、日历、地图、浏览器、联系人管理程序等。所有的应用程序都是用 Java 编写的。

② 应用程序框架。开发者完全可以访问核心应用程序所使用的 API 框架。该应用程序框架用来简化组件的重用，任何一个应用程序都可以发布它的功能块，并且任何其他的应用程序都可以使用其所发布的功能块（不过得遵循框架的安全性限制）。该应用程序重用机制使得组件可以被用户替换。

③ Android 程序库。Android 包括一个被 Android 系统中各种不同组件所使用的 C/C++ 集库。该库通过 Android 应用程序框架为开发者提供服务。

以下是一些主要的核心库。
- 系统 C 库：一个从 BSD 继承来的标准 C 系统函数库（Libc），专门为基于 Embedded Linux 的设备定制。
- 媒体库：基于 PacketVideo OpenCORE；该库支持录放，并且可以录制许多流行的音频及视频格式文件，播放静态映像文件，包括 MPEG4、H.264、AAC、JPG、PNG。
- Surface Manager：对显示子系统进行管理，并且为多个应用程序提供 2D 和 3D 图层的无缝融合。
- LibWebCore：一个最新的 Web 浏览器引擎，用来支持 Android 浏览器和一个可嵌入的 Web 视图。
- SGL：一个内置的 2D 图形引擎。
- 3D libraries：基于 OpenGL ES 1.0 APIs 实现，该库可以使用硬件 3D 加速（如果可用）或者使用高度优化的 3D 软加速。
- FreeType：位图（Bitmap）和向量（Vector）字体显示。
- SQLite：一个对于所有应用程序可用、功能强劲的轻型关系型数据库引擎。

④ Android 运行库。

Android 运行库提供了 Java 编程语言核心库的大多数功能。

每一个 Android 应用程序都在它自己的进程中运行，都拥有一个独立的 Dalvik 虚拟机实例。Dalvik 是基于同时高效地运行多个 VMs 实现的。Dalvik 虚拟机执行.dex 的 Dalvik 可执行文件，该格式文件对最小内存的使用做了优化。该虚拟机是基于寄存器的，所有的类都是经由 Java 汇编器编译，然后通过 SDK 中的 DX 工具转化成.dex 格式文件后由虚拟机执行。

Dalvik 虚拟机依赖于 Linux 的一些功能，比如线程机制和底层内存管理机制。

⑤ Linux 内核。Android 的核心系统服务依赖于 Linux 内核，如安全性保证、内存管理、进程管理、网络协议栈和驱动模型。Linux 内核也同时作为硬件和软件栈之间的硬件抽象层。

任务 2 搭建 Android 开发平台 IDE

【任务描述】

Eclipse 是一款基于插件式的绿色开源软件，可以将 ADT 插件集成到 Eclipse 环境下，基于 SDK 实现 Android 应用程序的开发。

具体步骤如下所示：

1）下载及安装 Eclipse；

2）下载及安装 Android SDK 开发工具包；

3）为 Eclipse 安装插件。

【任务实施】

（1）下载及安装 Eclipse

Eclipse 是当前比较流行的集成开发环境，通过安装不同的插件，可以支持各种类型的应用开发。使用 Eclipe 能够更快速地熟悉 Android 应用开发。用户可以通过 http://www.eclipse.org 下载并安装文件 Eclipse – jee – juno – SR1，选择 4.2 版本即可。下载后，只要将压缩文件解压即可，如解压到 C 盘根目录，如图 1-12 所示。

（2）下载及安装 Android SDK

SDK 是 Android 应用开发工具包，登录 http://www.androiddevtools.cn/站点可下载 4.2 版本。进入主页后单击菜单 "Android SDK Tools"，在弹出的下拉菜单中单击 "SDK" 进入如图 1-13 所示的界面。

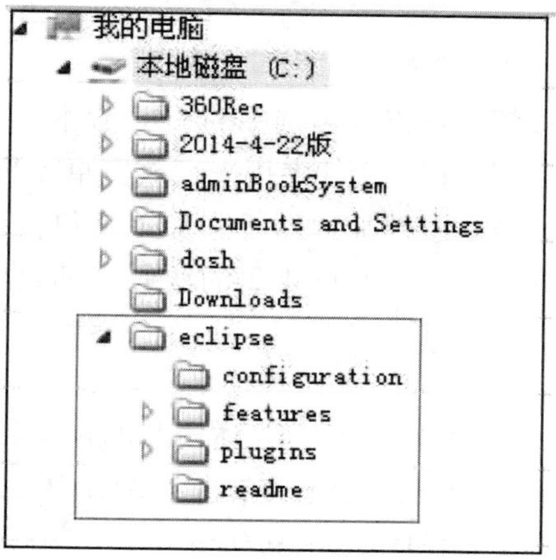

系统版本号	Windows	Mac OSX	Linux
android 5.0	下载	下载	下载
android L Rev3	下载	下载	下载
android L	下载	下载	下载
android 4.4W	下载	下载	下载
android 4.4.2	下载	下载	下载
android 4.3	下载	下载	下载
android 4.2.2	下载	下载	下载
android 4.1.2	下载	下载	下载
android 4.0.3	下载	下载	下载
android 4.0	下载	下载	下载
android 3.2	下载	下载	下载
android 3.1	下载	下载	下载
android 3.0	下载	下载	下载
android 2.3.3	下载	下载	下载
android 2.2	下载	下载	下载

图 1-12　Eclipse 安装目录

图 1-13　Android SDK 下载的链接

1) 单击"windows"列中的"Android4.2.2"链接，即可下载安卓4.2SDK压缩包。

2) 下载后，将文件解压到C盘根目录，将得到一个android-sdk-windows文件夹，此文件夹下包括3个子文件夹及两个.exe文件。

① add-ons 文件夹：该目录用来存放第三方公司为安卓平台开发的附加功能，刚解压后，此文件夹为空。

② platforms 文件夹：该目录用来存放不同版本的安卓平台，刚解压后，此目录也为空。当安装完成后，此文件夹将有不同版本的安卓平台。

③ tools 文件夹：此目录下存放的是关于安卓开发及调试的一些工具，如sqlite3、emulator、ddms 等。

④ SDK Manager.exe：该程序是安卓SDK管理器，通过这个工具可以管理安卓SDK。

⑤ AVD Manager.exe：该程序为安卓虚拟设备管理器，通过该工具可以管理AVD。

3) 双击SDK Manager.exe，可以看到如图1-14所示的窗口。

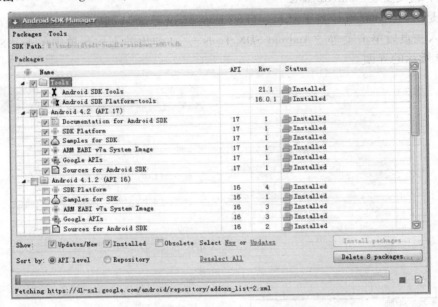

图1-14　安卓SDK及AVD管理器

勾选上图左边的复选框，单击Install packages按钮，可以安装选中的SDK版本。当安装完成后，可以看到整个SDK目录下增加了几个文件，如docs文件夹用来存放安卓SDK开发文件和相应的API文档等；extras文件夹存放一些驱动及硬件加速器等附加工具包；samples文件夹用来存放一些示例程序代码；sources文件夹用来存放安卓4.2的源码。其具体目录结构如图1-15所示。

（3）为Eclipse安装ADT插件

如果要在Eclipse中进行Android开发，还需要安装ADT插件，具体安装的步骤如下。

1) 登录http：//www.androiddevtools.cn/，在弹出的下拉菜单中单击"ADT Plugin"后即可下载ADT插件，本书中采用的版本是ADT版本，即ADT-21.0.0zip。

2) 双击图1-12中解压后的文件eclipse.exe，运行Eclipse后，在Eclipse运行界面的主菜单中选择 Help→Install New Software 菜单命令，如图 1-16 所示。选中上述菜单命令后，会弹出如图 1-17 所示的对话框。

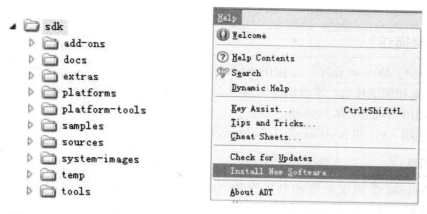

图 1-15　SDK 目录结构　　　　　　　图 1-16　选择菜单命令

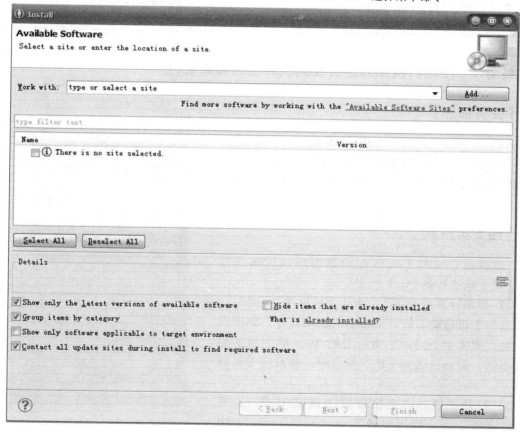

图 1-17　插件安装对话框

3）单击上图中的 Add 按钮，在弹出的对话框中单击 Archive 按钮，选中下载的插件 ADT-21.0.0.zip 后单击 OK 按钮，即可按照提示安装完成 ADT 插件。

经过上面的步骤，接下来就可以在 Eclipse 环境中开发 Android 应用了。

模块二　安装、运行及调试环境

【模块描述】

要想运行 Android 程序，必须在 Android 手机上才可以。因此，Android 开发必须准备相关运行环境和调试环境，本模块介绍两种 Android 程序运行和调试环境方法。

1）使用真机进行运行调试。
2）使用 AVD 即 Android 虚拟机进行运行调试。

知识点	技能点
➢ Android 虚拟模拟器和真机作为 APP 运行调试环境的区别及优缺点 ➢ Android APP 应用结构 ➢ Android SDK 版本及核心 API	➢ Android 真机作为运行及调试环境的参数设置 ➢ AVD 虚拟机的参数设置及启动

任务 1　使用真机作为运行及调试环境的参数配置

【任务描述】

APP 应用程序必须在 Android 手机上才可以运行，本任务为读者介绍使用 Android 真机作为 APP 应用程序运行及调试的环境。真机调试具有速度快、效率高的优点。

【任务实施】

1）准备一根和手机匹配的数据线（USB 连接线），将手机连接到 PC 上。

2）如果计算机安装了 "360 手机助手"，此时计算机会为手机自动下载并安装手机驱动程序。如果没有安装，则由于不同的手机厂商的 Android 手机的驱动都不同，需要自己到手机厂商官网下载对应手机的驱动。

3）打开手机，将手机设置为调试模式。步骤为："设置"→"应用程序"→"开发"→"USB 调试"，勾选"保持唤醒状态""允许模仿位置"及"USB 调试"3 个复选框即可，分别如图 1-18、图 1-19、图 1-20 所示。此时需要注意的是，小米手机尤其是红米手机，在设置的"开发者选项"中，需要首先开启"开发者选项"。

图 1-18　设置界面

图 1-19 开发选项

图 1-20 勾选复选框

完成如上步骤,就可以使用 Android 真机进行应用程序的调试工作了。

任务 2　使用 AVD 作为运行及调试环境的参数配置

【任务描述】

除了使用真机外,Android SDK 还为开发者提供了可以在计算机上运行的"虚拟手机",即 Android Virtual Device,简称 AVD。基于 AVD 调试 APP 应用程序,为没有 Android 智能手机的开发者带来了便利。

【任务实施】

1)通过运行 Android SDK 安装目录下的 AVD Manager.exe 可以启动 AVD 管理器,如图 1-21 所示。

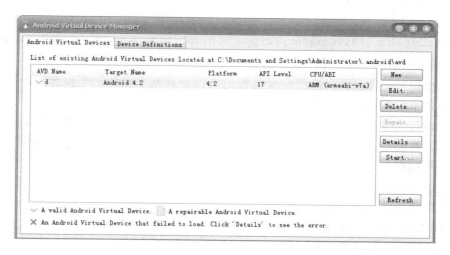

图 1-21 AVD 管理器

2)在上图中已经看到,创建了一个名称为 d 的模拟器,此模拟器是基于 Android 4.2 版本的。如果列表中没有模拟器，则可以单击 New 按钮创建一个模拟器，如图 1-22 所示。

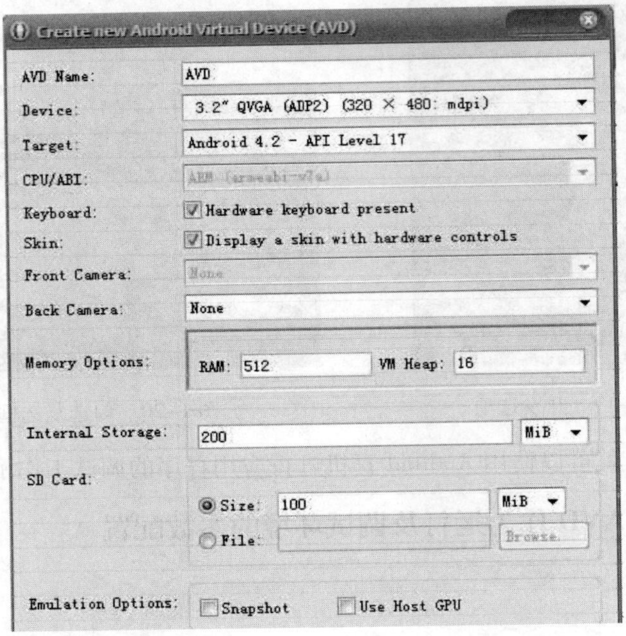

图 1-22　创建 AVD

图 1-22 中，AVD Name 为模拟器设备的名字；Device 为选择的所需开发屏幕的分辨率；Target 为选择的 Android 平台的版本，这里选择的是 4.2 版本；RAM 指的是设置手机的内存大小；VM Heap 指虚拟手机堆内存的大小；Size 指的是手机 SD 卡的大小。

如果想修改某个已经创建的模拟器，则选中模拟器，然后单击图 1-21 中的 Edit 按钮即可进入修改页面，也可以单击 Delete 按钮删除模拟器。

当手机模拟器创建完成后，就可以使用模拟器进行安卓应用程序的运行及调试了，具体步骤如下。

选中图 1-21 中的已经创建好的模拟器。单击右边的 Start 按钮，即可启动虚拟手机了，如图 1-23 所示。

上图的虚拟手机界面和自己的安卓真机的操作方式基本相似。当包含的应用程序较多时，可以通过手指左右拖动来查看更多的程序。

其实刚运行的时候，界面是英文的，对于国内用户来说，设置中文操作界面更符合操作习惯。在英文界面下，用户可通过 setting→language ＄input→language 进行设置中文操作界面，语言界面如图 1-24 所示。

【练习】

1. 搭建 Android 开发环境。
2. 创建一个 Android 手机模拟器。
3. 在 PC 上配置 Android 真机，调试并运行 Android 程序。

图 1-23 虚拟手机

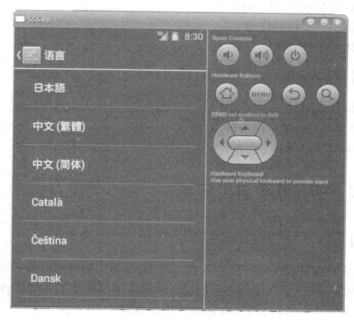

图 1-24 语言界面

项目二 开始我的第一个 Android 应用程序

随着移动互联网的迅猛发展，Android 平台 APP 应用开发越来越受开发者的追捧。对于初学者来说，Android 应用 APP 开发不知如何下手。对于这种情况，本部分将给出在 ADT 平台下开发并运行第一个 Android 应用程序的方法和步骤，并在分析 Android 应用程序结构的基础上重点介绍使用 DDMS 对应用程序进行调试的步骤和注意事项，最后给出了在 Eclipse 中对 Android 应用进行签名及打包资源的方法。

【知识目标】

- 基于 Android Eclipse IDE 开发的方法和步骤
- Android 应用程序结构
- 使用 DDMS 对 Android 应用进行调试
- 资源文件在 Android 应用中的作用
- Android 全局描述文件的作用
- Android 应用程序的签名

【模块分解】

- 使用 ADT 创建 Android 应用工程
- 使用 DDMS 对 Android 应用进行调试
- 使用 Eclipse 对 Android 应用进行签名

模块一 创建 Android 应用工程

【模块描述】

Android 应用开发相对容易，只要读者有 Java SE 桌面应用程序开发的经历或经验就不难掌握；相对于开发 Java SE 桌面应用程序的读者来说，仅仅增加了一些 Android API 罢了。

使用 Eclipse 开发 Android 应用大致有以下几个步骤：
1）创建一个安卓工程和应用程序；
2）在 XML 布局文件中定义用户界面；
3）编写 Java 代码以实现相关的业务功能；
4）创建、启动 AVD 并调试 APP；
5）APP 签名、打包发布。

知识点	技能点
➢ Android 开发平台及开发环境 ➢ Android APP 应用结构 ➢ Android 相关资源分类及使用	➢ 基于 ADT 搭建 APP 项目的方法和步骤 ➢ 资源文件的简单使用 ➢ 按照项目需求搭建 APP 应用程序结构

任务 1 基于 ADT 新建 Android 项目

【任务描述】

在 ADT 平台下可视化创建 Android APP 应用时，指定应用的图标、Android 应用名、Android 项目名、Android 应用程序包、运行的最低版本要求、Activity 名称、布局文件名称等。

【任务实施】

1）新建项目，选择 File→new→Project 菜单命令，在弹出的对话框中选择 Android Application Project 选项，如图 2-1 所示。

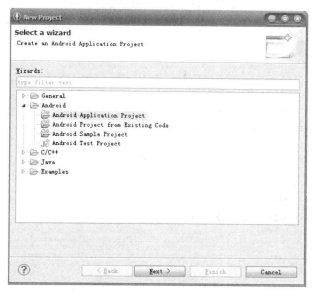

图 2-1 新建项目方式一

或者按照图 2-2 所示，选中 File→New→Android Application Project 菜单命令。

2）弹出如下对话框，在此对话框中输入或选择一些项目信息，如图 2-3 所示。

① Application Name：表示 Android 应用程序的名称，这个是开发出来的程序安装到设备中之后在图标下面显示的名字。

② Project Name：表示建立的 Android 项目名称。

③ Package Name：包的名称，这个相当于 .Net 中的 Namespace。

④ Minimum Required SDK：指定运行 Android 应用程序的最低版本要求。

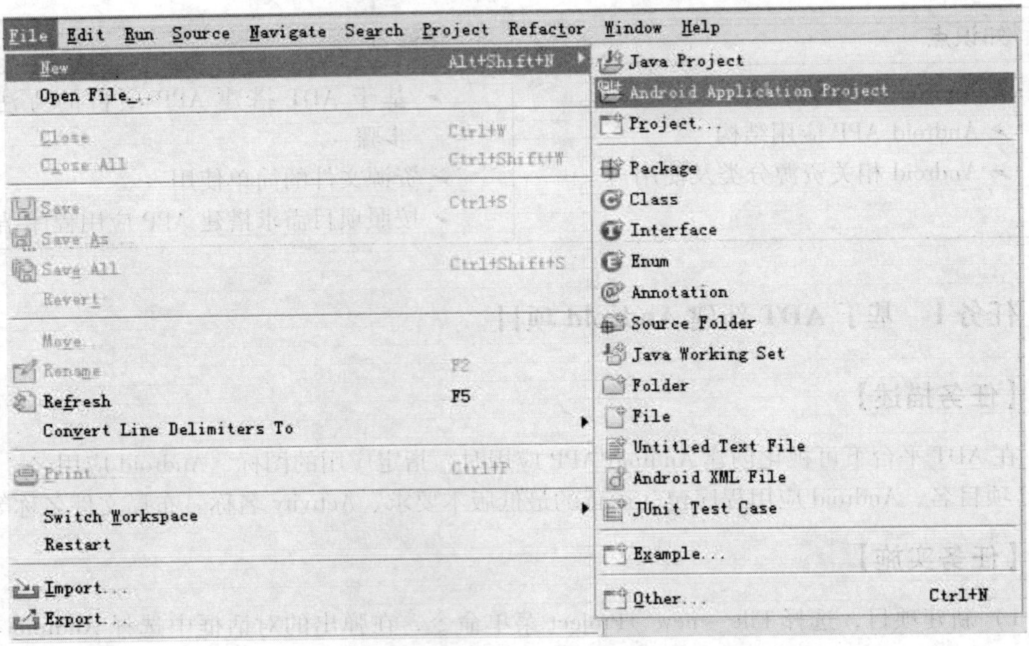

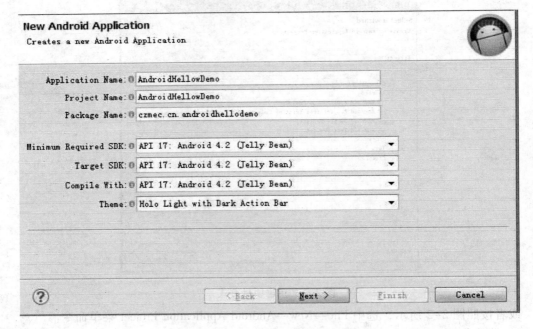

图 2-2　新建工程方式二

图 2-3　设置项目信息

⑤ Target SDK：说明该应用程序对应的那个 Android 版本。

3）完成上述设置后，单击 next 按钮弹出如图 2-4 所示的对话框。

4）如果没有在上图中勾选 Create custom launcher icon 复选框，应用将会采用 Android SDK 默认的图标；如果勾选了这个复选框，则系统会弹出如图 2-5 所示的对话框，在这个对话框中可以设置文字或图标。

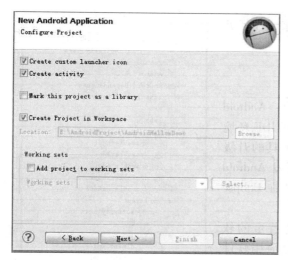

图 2-4　配置项目对话框

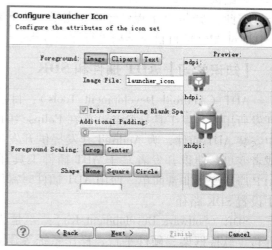

图 2-5　制作自定义图标对话框

其中，Image：使用图片制作图标；Clipart：使用系统图片制作图标；③Text：使用文字制作图标；④Shape：可以选择图标的形状，Square 为方形，Circle 为圆形。

5）单击上图的 Next 按钮，进入如图 2-6 所示的对话框。

6）如果不勾选上图中的 Create Activity，可以直接单击 Finish 按钮完成项目的创建。如果勾选了这个复选框，则单击 Next 按钮，进入如图 2-7 所示的创建 Activity 的对话框，其中，列表中的内容为 Activity 的模板。

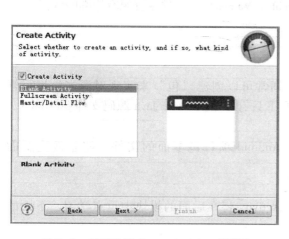

图 2-6　是否设置 Create Activity 对话框

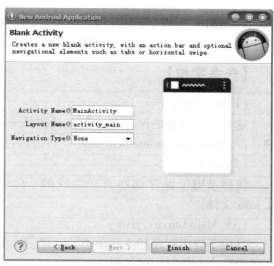

图 2-7　创建 Activity 对话框

图 2-7 中的 Activity Name 为创建的 Activity 的名称，Layout Name 为界面布局文件的名称，它是一个 .xml 文件。

单击 Finish 按钮，将成功创建一个 Android 项目，其目录结构如图 2-8 所示。

至此，Android 工程已经建立完毕，接下来，在任务 2 中将对 APP 项目目录结构进行详细分析。

【知识学习】ADT 插件和 SDK

ADT（Android Development Tools）：目前，Android 开发所用的开发工具是 Eclipse，在 Eclipse 编译 IDE 环境中安装 ADT 插件，为 Android 开发提供开发工具的升级或者变更，使得开发者基于 ADT 插件工具开发 Android APP 应用变得非常简单。使用 ADT 插件时需要在 Eclipse 中设置 SDK 路径。

SDK（Software Development Kit）：一是为特定的软件包、软件框架、硬件平台、操作系统等建立应用软件的开发工具的集合。SDK 是专门为开发者开发 Android 应用提供了库文件及其他开发所用到的工具。用户可简单理解为开发工具包集合，是整体开发中所用到的工具包，如果不用 Eclipse 作为开发工具，就不需要下载 ADT，只下载 SDK 即可开发。

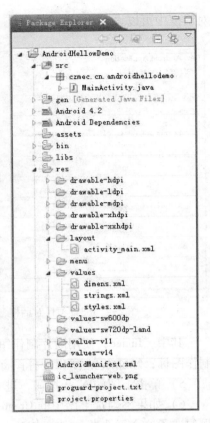

图 2-8　Android 项目目录结构

任务 2　Android 应用程序结构分析

【任务描述】

Android 应用程序结构比较复杂，尤其是众多资源文件的引用，初学者需要对 APP 应用结构中的 src、gen、activity、res（Drawables、Layouts、Values）等文件或资源有详细的认识。

【任务实施】

（1）src 文件夹

这是一个普通的保存 Java 源文件的目录。通常可以创建"包"来实现对源文件进行管理，即 czmec.cn.androidhellodemo。目前该目录下只有一个 Activity Java 源码文件。

（2）gen 文件夹

该目录用于保存那些自动生成的、位于 Android 项目包下面的文件，如非常重要的 R.java 文件。

（3）MainAcitvity.java

其用于应用程序的主活动类的实现。这个 Activity 将在后面的模块中重点学习。下面的代码为这个文件对应的源码。

1.　import android.os.Bundle;
2.　import android.app.Activity;
3.　import android.view.Menu;

```
4.    public class MainActivity extends Activity {
5.        protected void onCreate(Bundle savedInstanceState) {
6.            super.onCreate(savedInstanceState);
7.            setContentView(R.layout.activity_main);
8.        }
9.        public boolean onCreateOptionsMenu(Menu menu) {
10.           // Inflate the menu; this adds items to the action bar if it is present.
11.           getMenuInflater().inflate(R.menu.main, menu);
12.           return true;
13.       }
14.   }
```

从代码中可以看到，MainActivity 继承于 Activity 类，Activity 是 Android 中的视图部分，负责处理界面显示。在 MainActivity 里面重写了父类的 onCreate 方法和 onCreateOptionsMenu 方法，在重写的 onCreate 方法里，方法 setContentView(R.layout.activity_main)给 MainActivity 设置了要显示的视图 R.layout.activity_main，视图由 R 类寻找并加载（就像 mvc，Activity 相当于 Controller，而要显示的 Layout 就相当于具体的页面）。

对于 Java 基础不错的读者来说，上面这个程序容易明白。

（4）Referenced libraries

其包含 android.jar，这是 Android SDK 中运行时类的 jar 文件。

（5）res 文件夹

这个目录非常重要，用来存放 Android 项目的各种资源文件，如布局文件、字符串资源文件、颜色资源文件、尺寸资源文件等，包括 Drawables、Layouts、Values 等子目录。

1）Drawables。这个文件夹包含图形文件，比如图标和位图，Drawable – xdpi（hdpi、ldpi、mdpi）用以设置对应不同的分辨率。hdpi 对应的是 800 像素 * 480 像素及以上，当应用程序安装到不同分辨率的机器上的时候，其会到对应的文件夹中去读取。

2）Layouts。这个文件夹包含表示应用程序布局和视图的 XML 文件。可以看到，这里已经有了一个 XML 文件 activity_main.xml。这个文件的内容如下。

```
1.    <RelativeLayout xmlns:android = "http://schemas.android.com/apk/res/android"
2.        xmlns:tools = "http://schemas.android.com/tools"
3.        android:layout_width = "match_parent"
4.        android:layout_height = "match_parent"
5.        android:paddingBottom = "@dimen/activity_vertical_margin"
6.        android:paddingLeft = "@dimen/activity_horizontal_margin"
7.        android:paddingRight = "@dimen/activity_horizontal_margin"
8.        android:paddingTop = "@dimen/activity_vertical_margin"
9.        tools:context = ".MainActivity" >
10.       <TextView
11.           android:id = "@+id/hello"
12.           android:layout_width = "wrap_content"
```

```
13.            android:layout_height = "wrap_content"
14.            android:text = "@ string/hello_world"/>
15.    </RelativeLayout>
```

可以看到,这个 XML 文件中定义了一个根节点"RelativeLayout"和一个子节点"TextView"。其中,根节点指定了运行界面的布局方式——相对布局,子节点则定义了一个 ID 为"hello"的文本控件,同时指定了其一些诸如高度、宽带等的属性值。在后面的模块中将会重点介绍。

3) Values。这个文件夹包含 strings.xml 文件等。这是应用程序实现字符串本地化的主要方法。UI 遵循了 MVC 开发模式,UI 上用到的字符名称都可以定义在这个文件夹下。这个文件的内容如下。

```
1.  <? xml version = "1.0" encoding = "utf-8"?>
2.  <resources>
3.      <string name = "app_name">AndroidHellowDemo</string>
4.      <string name = "action_settings">Settings</string>
5.      <string name = "hello_world">Hello World!</string>
6.  </resources>
```

上述代码的第 5 行指定了运行界面中显示的内容。其中,<string name = "hello_world">中指定的"hello_world"即为上述 activity_main.xml 文件中第 14 行对应的"hello_world",即 android:text = "@ string/hello_world"。

关于 string 字符串资源的使用,在以后的模块中也将会重点介绍。

(6) R.java

Android Developer Tools 自动创建这个文件,是由 AAPT 工具根据应用过程中的各种资源文件自动生成的,它提供访问 Android 应用程序的各种资源所需的常量。后面会详细讨论 R 类与资源之间的关系。下面的代码为这个文件对应的源码。

```
1.  public final class R {
2.      public static final class attr {
3.      }
4.      public static final class dimen {
5.          public static final int activity_horizontal_margin = 0x7f040000;
6.          public static final int activity_vertical_margin = 0x7f040001;
7.      }
8.      public static final class drawable {
9.          public static final int ic_launcher = 0x7f020000;
10.     }
11.     public static final class id {
12.         public static final int action_settings = 0x7f080000;
13.         public static final int hello = 0x7f080000;
14.     }
```

```
15.     public static final class layout {
16.         public static final int activity_main = 0x7f030000;
17.     }
18.     public static final class menu {
19.         public static final int main = 0x7f070000;
20.     }
21.     public static final class string {
22.         public static final int action_settings = 0x7f050001;
23.         public static final int app_name = 0x7f050000;
24.         public static final int hello_world = 0x7f050002;
25.     }
26.     public static final class style {
27.         public static final int AppBaseTheme = 0x7f060000;
28.         public static final int AppTheme = 0x7f060001;
29.     }
30. }
```

此文件是在构建时自动创建的,所以不要手工修改它,因为所有修改都会丢失。可以把 R.java 理解成 Android 应用的资源字典。

AAPT 工具在生成 R.java 文件时,其规则如下。

1) 每类资源对应 R 类中的一个内部类。如字符串资源对应 string 内部类,所有标识符资源对应于 id 内部类。

2) 每个具体的资源项对应于内部类的一个 public static final int 类型的 field 域。如在应用的界面上有一个名称为"hello"的文本框,则在 R.java 中的内部类中有一个对应的"hello"的 int 类型的常量。

随着 Android 项目中的资源越来越多,R.java 文件中的相关内容也会越来越多。

(7) bin 目录

该目录用来存放生成的目标文件,如 Java 的字节码文件、资源打包文件和 Dalvik 虚拟机的可执行文件(.dex 文件)。

(8) AndriodManifest.xml

这是应用程序的部署描述符文件,里面定义了 Android 项目的系统清单文件,它除了可定义 Android 应用程序的名称、图标及访问权限等属性外,还可以定义 Android 应用的 Activity、Service 等组件。

任务3 资源文件的使用方式

【任务描述】

Android 应用 APP 中需要用到大量的资源,包括字符串资源、颜色资源、尺寸资源、数组资源、图片资源、布局资源、菜单资源、Drawable 资源、XML 资源等,APP 应用的编程很大程度上就是对各种资源进行引用的过程。因此本任务分析资源的两种常见使用方式,在后面任务中详细介绍。

【任务实施】

(1) 在 Java 文件中使用资源

为了在 Java 代码中使用资源，SPK 的 build – tools 目录下的 AAPT（Andriod Asset Packaging Tool）会为 Android 项目自动生成一份 R. java 文件，其中，每个资源项对应内部类里面的一个 int 类型的域。如 string. xml 资源文件中定义的"app_name"字符串常量，即"< string name = "app_name" > Hello World < /string >"，在 R. java 文件中有一个对应的常量：

```
public static final int app_name = ox7f040000;
```

那么，在 Java 代码中就可以使用"R. string. app_name"来引用"Hello World"字符串常量。

同样，在 MainActivity 类代码中使用"setContentView(R. layout. activity_main)"方法指定了要显示的视图，其中，"R. layout. activity_main"也使用同样的方法。

(2) 在 XML 文件中使用资源

在 XML 文件中使用资源更加简单，只要按照如下的格式访问即可：

```
@ <资源对应的内部类的类名 >/<资源项名称 >
```

如要访问上述字符串资源中定义的"hello World"字符串常量，则可以使用如下方式引用：

```
@ string/app_name
```

但也有一种例外情况，当在 XML 文件中使用标识符时，这些标识符无须使用专门的资源进行定义，直接在 XML 文档中按照如下格式分配标识符即可：

```
@ + id/ < 标识符代号 >
```

如上面的案例中定义的一个文本组件：

```
android:id = "@ + id/text"
```

上面的代码为这个文本组件分配了一个标识符，这个标识符在 Java 代码中可以直接使用，即可以使用 Activity 类中的 findViewById( )用以获取该组件；如果要在 XML 文件中获取这个组件，可以通过资源引用的方式直接引用它，即：

```
@ + id/text
```

模块二 运行及调试 APP 应用

【模块描述】

Android 的 APP 应用可以在真机和虚拟设备 AVD 下进行调试，用户需要配置 AVD，启动 AVD，再在 AVD 下运行并调试 APP 应用。接下来详细介绍使用 ADT 运行 Android APP 应用的步骤。

知识点	技能点
➢ Android 虚拟设备 AVD 的配置 ➢ Android 应用运行、调试方式	➢ AVD 配置、启动方法 ➢ APP 应用运行调试步骤

任务 1　启动 AVD

【任务描述】

AVD 就是指 Android 模拟器。在运行 APP 应用之前，需要启动 AVD。

【任务实施】

通过在 Eclipse 中选择 Window→Android Virtual Device Manager 菜单命令来启动，如图 2-9 所示。

或者直接单击工具栏中的"Android Virtual Device Manager"按钮，可以弹出如图 2-10 所示的对话框，单击图中的 Start 按钮就可以启动 AVD。

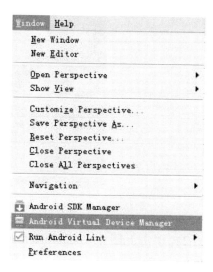

图 2-9　选择菜单命令

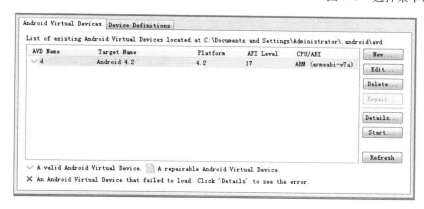

图 2-10　AVD 启动

任务 2　运行 APP

【任务描述】

在 ADT 平台下运行 APP 应用，并注意运行时的问题分析。

【任务实施】

1）在 Project Explorer 中右击项目名，在弹出的快捷菜单中选择 Run As→Android Application 命令，就可以运行应用程序了，如图 2-11 所示。

图 2-11　程序运行（一）

2）单击上图中 AndroidHellow Demo 的，即可以看到程序运行的界面了，如图 2-12 所示。

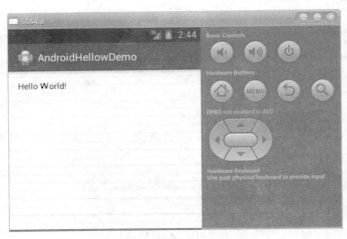

图 2-12　程序运行（二）

使用模拟器的时候有以下几点需要注意：
① 模拟器启动时间有点长，所以电脑启动后模拟器的开启需要时间较长，具体时间取决于开发用计算机的配置。
② 第一遍运行完成之后不需要急着关闭模拟器，第二次运行程序时只要单击 Run 按钮就会自动把新的程序安装到模拟器，然后自动执行（以 Log 为证）。

任务 3　使用 DDMS 进行调试

【任务描述】

APP 运行遇到错误时,需要进行调试才能查到问题所在,基于 DDMS 完成 APP 应用的调试任务,领会调试的意义和技巧。

【任务实施】

1)打开 MainActivity.java 类(本项目模块—任务 1 创建的项目),添加如下的 Log 记录功能。

```
1.  public class MainActivity extends Activity {
2.      protected void onCreate(Bundle savedInstanceState) {
3.          Log.d("LOG_CAT","程序开始启动...");
4.          super.onCreate(savedInstanceState);
5.          setContentView(R.layout.activity_main);
6.          Log.d("LOG_CAT","程序运行结束");
7.      }
8.  }
```

2)按〈Ctrl + S〉组合键保存刚才的修改,重新运行应用程序,打开 DDMS(Dalvik Debug Monitor Service,Dalvik 虚拟机调试监控服务),找到 LogCat 窗口,在搜索栏输入"程序",就可以看到刚才写的 Log 信息。

LogCat 窗口如图 2-13 所示。

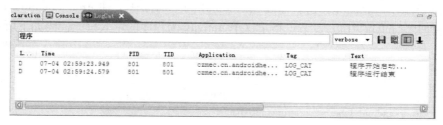

图 2-13　LogCat 窗口

至此,用 Eclipse 开发 Android 应用程序已经介绍完了,包括 Android 认识、Android 开发环境搭建、Android 程序创建、运行、简单调试。

【知识学习】Activity 组件概述

Activity 是 Android 应用的重要组成单元之一,并且是非常常用的组件之一。Activity 是 Android 应用的入口,是初级程序员必须掌握的组件之一。

从上面的案例可以看出,创建一个 Android 应用必须至少有一个 Activity 类,如上述类 MainAcitvity 就是继承了 Activity 的一个类。

对于初级开发者而言,到底该如何理解 Activity 的本质呢?

1)从表面来看,Activity 是 Android 应用的一个图形用户界面。

2)任何一个 Android 应用都至少有一个 Activity 类。

3）对于开发者来说，Activity 就是程序的入口，是一个有一定编程规范的 Java 类。

从应用角度看，Activity 是一个 Java 类，该类符合一定的规范，在定义这个 Activity 类时，可以直接或间接地继承 android.app.Activity 类，定义的这个类存放于 Android 应用程序下的 src 目录下，如上述的第一个 Android 应用程序。

Android 应用中并没有类似 Java 应用那样的 main 方法，Activity 即是 Android 应用运行的入口，如上面的案例中定义的 MainAcitvity 类的代码：

```
1.  public class MainActivity extends Activity {
2.      protected void onCreate(Bundle savedInstanceState) {
3.          super.onCreate(savedInstanceState);
4.          setContentView(R.layout.activity_main);
5.      }
6.      public boolean onCreateOptionsMenu(Menu menu) {
7.          // Inflate the menu; this adds items to the action bar if it is present.
8.          getMenuInflater().inflate(R.menu.main, menu);
9.          return true;
10.     }
11. }
```

可见，MainAcitvity 继承了 android.app.Activity 类，并覆盖了其中的 onCreate 方法。当运行这个程序时，将自动调用该类的 onCreate 方法，构建应用的图形用户界面。

Activity 创建完成后不能直接被调用运行，还必须进行配置。同样，一个 Android 应用不可能只有一个界面，那就意味着不可能只有一个 Activity。那么，Android 应用如何知道先启动哪个 Activity 来显示用户界面呢？这就需要在 Android 的全局描述文件 AndroidMainFest.xml 中进行配置，并使用 <intent-filter> 节点指定这个 Activity 为程序的主入口，代码如下所示：

```
1.  …
2.  <activity
3.      android:name="czmec.cn.androidhellodemo.MainActivity"
4.      android:label="@string/app_name" >
5.      <intent-filter>
6.          <action android:name="android.intent.action.MAIN"/>
7.          <category android:name="android.intent.category.LAUNCHER"/>
8.      </intent-filter>
9.  </activity>
10. …
```

当然，android.app.Activity 类中定义了大量的常用方法和常量，这将在后面详细介绍。

模块三　签名并打包 Android 应用程序

【模块描述】

Android 应用的 APP 项目是以包名作为唯一标志，在一台手机上安装包名相同的应用时，后安装的应用会覆盖前面的应用，为避免此类情况，需要对 APP 应用进行签名。本模块主要实现应用签名及打包任务。

知识点	技能点
➢ APP 签名及数字证书 ➢ APP 应用程序的发布	➢ 在 Eclipse 平台下实现对 APP 应用签名

任务　APP 签名及打包

【任务描述】

对 Android 应用程序进行签名有两种方法，即使用 Eclipse 工具可视化签名方式和控制台中输入命令操作的方式，本书主要讲解采用第一种方式为 APK 进行签名的方法。

【任务实施】

使用 Eclipse 工具进行签名的方法适用于 Android 1.5 及以上版本。

1）打开 Eclipse，选择要签名的项目后右击，在弹出的快捷菜单中选择 Android Tools→Export Signed Application Package 命令，如图 2-14 所示。

如图 2-15 所示，在出现的对话框中确认是不是这个项目要签名，然后单击 Next 按钮。

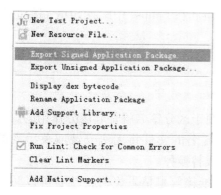

图 2-14　选择快捷菜单命令

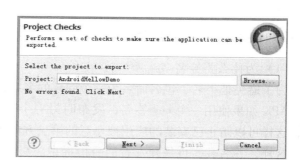

图 2-15　确定要签名的项目

2）在 Keystore selection 对话框，如果之前已有了 Keystore，选择之前已有的，否则新建一个。

如图 2-16 所示，选择需要保存这个证书文件的目录，以及设置这个证书文件的一个密码。

3）单击 Next 按钮后需要填写 Keystore 的基本信息，如别名、密码、有效期、姓名等，（图 2-17），之后单击 Next 按钮。

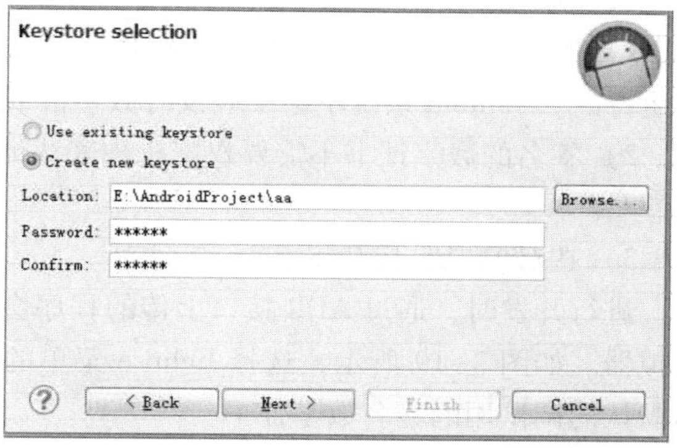

图 2-16　设置证书目录及密码

4)选择被签名后的 APK 并保存位置,然后单击 Finish 按钮,如图 2-18 所示。注意是选择最终产生的文件。

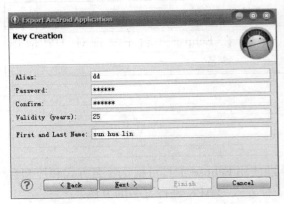

图 2-17 设置签名基本信息

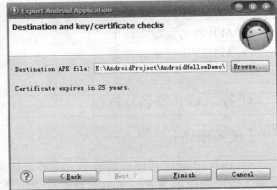

图 2-18 设置保存位置

之后在刚才选择的目录下就可以看到生成的签名后的 APK 文件。

签名打包注意事项如下。

① SDK 的安装目录不要有中文和空格,否则会出现不必要的错误。

② 如果 ADT 安装路径包含 Program Files(有空格),用 ADT 打包会报错,用 Ant 方式打包时也会报错。此时的解决方法:配置 SDK 环境变量和在 Eclipse 中配置 SDK 路径的时候把 Program Files 改为 Progra～1 即可。

【知识学习】签名注意事项

(1)为什么要签名

为了避免把类名、包名命名为一个同样的名字,便于区分签名就是起区分作用的。

APK 如果使用一个 key 签名,发布时另一个 key 签名的文件将无法安装或覆盖老的版本,这样可以防止已安装的应用被恶意的第三方覆盖或替换掉。

这样,签名其实也是开发者的身份标识。交易中抵赖等事情发生时,签名可以防止其发生。

(2)签名的注意事项

1)所有的 Android 应用都必须有数字签名,没有不存在数字签名的应用,包括模拟器上运行的。Android 系统不会安装没有数字证书的应用。

2)签名的数字证书不需要权威机构来认证,是开发者自己产生的数字证书,即所谓的自签名。

3)在对模拟器开发环境

进行开发时,通过 ADB 接口上传的程序会先自动被签有 Debug 权限,然后才被传递到模拟器。如图 2-19 所示,选择 Eclipse 菜单的 Window→Preferences→Android→Build,会显示默认的调试用的签名数字证书。

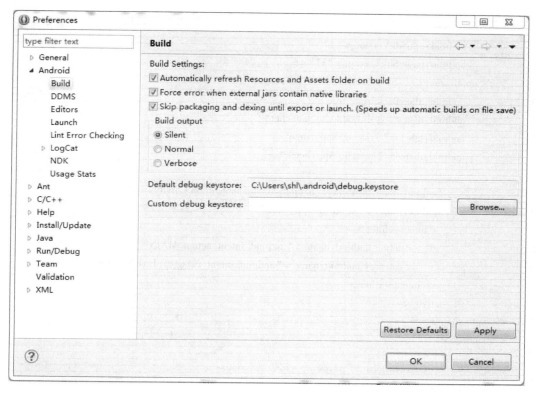

图 2-19 签名数字证书

4）正式发布一个 Android 应用时，必须使用一个合适的私钥生成的数字证书来给程序签名，不能使用 ADT 插件或者 ANT 工具生成的调试证书来发布。

5）数字证书都是有有效期的，Android 只是在应用程序安装的时候才会检查证书的有效期。如果程序已经安装在系统中，那么即使证书过期也不会影响程序的正常功能。

6）数字证书用来标识应用程序的作者和应用程序之间的信任关系，而不是用来决定最终用户可以安装哪些应用程序。

【问题研讨】 AndroidManifest 文件

这是个 XML 文件，它是应用程序的配置文件，包含在每个 Android 应用程序中，它向系统描述了本程序所包括的组件、所实现的功能、所能处理的数据、要请求的资源等，可以近似看作网站中的 Web.config 文件。同样，它也可以由可视化编辑器或文本编辑器编辑，如图 2-20 所示。

对应的代码如下所示。

```
1.  <? xml version = "1.0" encoding = "utf-8"? >
2.  < manifest xmlns:android = "http://schemas.android.com/apk/res/android"
3.      package = "czmec.cn.androidhellodemo"
4.      android:versionCode = "1"
5.      android:versionName = "1.0" >
```

```
6.      < uses – sdk
7.          android:minSdkVersion = "17"
8.          android:targetSdkVersion = "17"/ >
9.      < application
10.         android:allowBackup = "true"
11.         android:icon = "@ drawable/ic_launcher"
12.         android:label = "@ string/app_name"
13.         android:theme = "@ style/AppTheme" >
14.         < activity
15.             android:name = "czmec.cn.androidhellodemo.MainActivity"
16.             android:label = "@ string/app_name" >
17.             < intent – filter >
18.                 < action android:name = "android.intent.action.MAIN"/ >
19.                 < category android:name = "android.intent.category.LAUNCHER"/ >
20.             < /intent – filter >
21.         < /activity >
22.     < /application >
23. < /manifest >
```

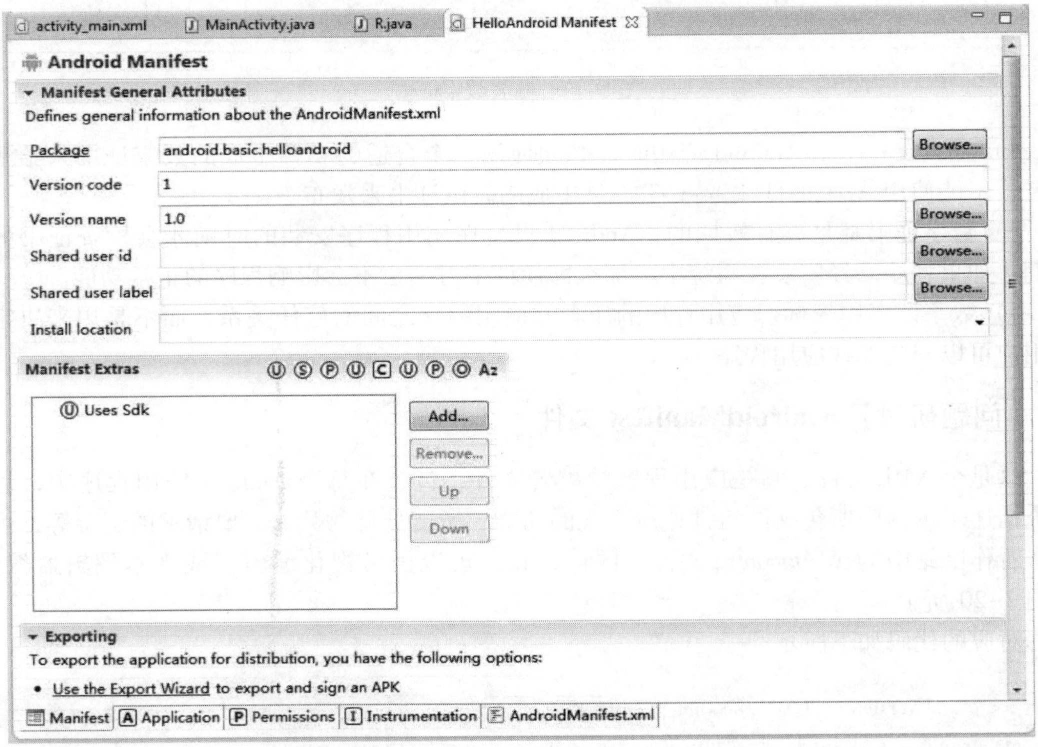

图 2-20 编辑器

第 3 行指定 Android 应用的包的名字，该包名可以用于唯一地标识该应用程序；第 11、

12、13 行指定 Android 应用的图标、标签和主题，ic_launcher、app_name 和 AppTheme 对应 string.xml 文件中相应的字符产值；第 14、15、16 行定义了一个 Activity 组件，该组件的具体实现类为 czmec.cn.androidhellodemo.MainActivity；第 17～20 行指定了程序的入口，当系统运行时将加载该 Acitvity。

【任务拓展】

一个 Android 应用可能调用另外一个 Android 应用程序，而这需要权限才能调用，因此需要在项目中声明调用自身所需要的权限。

（1）声明运行该应用程序所需的权限

可以在 <manifest> 节点中添加子节点 <uses-permission> 来声明。如下代码声明了该应用程序本身需要打电话的权限。

```
<users-permission android:name="android.permission.CALL_PHONE"/>
```

（2）声明调用该应用程序所需要的权限

通过为应用的各个组件元素如 <activity> 等元素添加 <uses-perission> 子节点，就可以声明调用该应用所需要的权限。

如在 <activity> 节点里面添加如下子节点：

```
<users-permission android:name="android.permission.SEND_SMS"/>
```

Android 提供了大量的权限，如表 2-1 所列。

表 2-1　Android 权限说明

权限	说明
ACCESS_COARSE_LOCATION	通过 WiFi 或移动基站的方式获取用户粗略的经纬度信息
ACCESS_FINE_LOCATION	通过 GPS 芯片接收卫星的定位信息，定位精度达 10 m 以内
ACCESS_MOCK_LOCATION	获取模拟定位信息，一般用于帮助开发者的调试应用
ACCESS_NETWORK_STATE	获取网络信息状态，如当前的网络连接是否有效
ACCESS_WIFI_STATE	允许程序访问 WiFi 网络状态信息
ACCOUNT_MANAGER	获取账户验证信息
ADD_SYSTEM_SERVICE	允许程序发布系统级服务
BATTERY_STATS	允许程序更新手机电池统计信息
BIND_DEVICE_ADMIN	获取系统管理权限
BIND_INPUT_METHOD	请求 InputMethodService 服务，只有系统才能使用
BIND_REMOTEVIEWS	必须通过 RemoteViewsService 服务来请求，只有系统才能用
BIND_WALLPAPER	必须通过 WallpaperService 服务来请求，只有系统才能用
BLUETOOTH	允许程序连接到已配对的蓝牙设备
BLUETOOTH_ADMIN	允许程序发现和与蓝牙设备配对
BRICK	能够禁用手机
BROADCAST_PACKAGE_REMOVED	允许程序，在一个应用程序包已经移除后广播一个提示消息

(续)

权限	说 明
BROADCAST_STICKY	允许一个程序广播常用 intents
CALL_PHONE	允许一个程序初始化一个电话拨号,不需要通过拨号
CALL_PRIVILEGED	允许一个程序拨打任何号码,包含紧急号码
CAMERA	请求访问使用照相设备
CHANGE_CONFIGURATION	允许一个程序修改当前设置,如本地化
CHANGE_NETWORK_ STATE	改变网络状态,如是否能联网
CHANGE_WIFI_MULTICAST_STATE	改变 WiFi 多播状态
CHANGE_WIFI_STATE	允许程序改变 WiFi 连接状态
CLEAR_APP_CACHE	清除应用缓存
CLEAR_APP_USER_DATA	清除应用的用户数据
CWJ_GROUP	允许 CWJ 账户组访问底层信息
CELL_PHONE_MASTER_EX	手机优化大师扩展权限
CONTROL_LOCATION_UPDATES	允许获得移动网络定位信息的改变
DELETE_CACHE_FILES	允许程序删除缓存文件
DELETE_PACKAGES	允许程序删除应用
DEVICE_POWER	允许访问底层电源管理
DIAGNOSTIC	允许程序 R/W(可读/写)诊断资源
DISABLE_KEYGUARD	允许程序禁用键盘锁
DUMP	允许程序从系统服务中获取系统 dump 信息
FACTORY_TEST	作为一个工厂测试程序,运行在 root 用户
FLASHLIGHT	允许访问闪光灯
FORCE_BACK	不管 Activity 是否在顶层都允许程序强制使用 back("后退")按键
GET_ACCOUNTS	访问 GMail 账户列表
GET_PACKAGE_SIZE	允许一个程序获取任何 package 占用空间容量
GLOBAL_SEARCH	允许程序使用全局搜索功能
HARDWARE_TEST	访问硬件辅助设备,用于硬件测试
INJECT_EVENTS	允许一个程序截获用户事件,如按键、触摸、轨迹球等
INSTALL_LOCATION_PROVIDER	定位提供的安装
INSTALL_PACKAGES	允许程序安装应用
INTERNAL_SYSTEM_WINDOW	允许程序打开内部窗口,不对第三方应用程序开放此权限
INTERNET	访问网络连接
MANAGE_ACCOUNTS	允许程序管理 AccountManager 中的账户列表
MANAGE_APP_TOKENS	允许应用去管理(创建、销毁)在窗口管理者中的应用
MTWEAK_USER	允许 mTweak 用户访问高级系统权限
MTWEAK_FORUM	允许使用 mTweak 社区权限
MASTER_CLEAR	允许程序执行软格式化,删除系统配置信息

(续)

权　限	说　明
MODIFY_AUDIO_SETTINGS	修改声音设置信息
MODIFY_PHONE_STATE	修改电话状态，如飞行模式，但不包含替换系统拨号器界面
MOUNT_FORMAT_FILESYSTEMS	格式化可移动文件系统，比如格式化清空 SD 卡
MOUNT_UNMOUNT_FILESYSTEMS	挂载、反挂载外部文件系统
NFC	允许程序执行 NFC 近距离通信操作，用于移动支持
PERSISTENT_ACTIVITY	创建一个永久的 Activity
PROCESS_OUTGOING_CALLS	允许程序监视、修改或放弃电话的拨出
READ_CALENDAR	允许程序读取用户的日程信息
READ_CONTACTS	允许应用访问联系人通讯录信息
READ_FRAME_BUFFER	读取帧缓存用于屏幕截图
READ_HISTORY_BOOKMARKS	读取浏览器收藏夹和历史记录
READ_INPUT_STATE	读取当前键的输入状态，仅用于系统
READ_LOGS	允许程序读取底层系统日志文件
READ_PHONE_STATE	访问电话状态
READ_OWNER_DATA	允许程序读取所有者数据
READ_SMS	允许程序读取短信息
.READ_SYNC_SETTINGS	读取同步设置，读取 Google 在线同步设置
READ_SYNC_STATS	读取同步状态，获得 Google 在线同步状态
REBOOT	允许程序重新启动设备
RECEIVE_BOOT_COMPLETED	允许程序开机自动运行
RECEIVE_MMS	接收彩信
RECEIVE_SMS	接收短信
RECEIVE_WAP_PUSH	接收 WAP PUSH 信息
RECORD_AUDIO	录制声音，通过手机或耳机的麦克进行录制
REORDER_TASKS	允许程序改变 Z 轴排列任务
SEND_SMS	发送短信
SET_ACTIVITY_WATCHER	允许程序监控或控制已经在全局系统中启动的 activites
SET_ALARM	设置闹铃提醒
SET_ALWAYS_FINISH	设置程序在后台是否总是退出
SET_ANIMATION_SCALE	设置全局动画缩放
SET_DEBUG_APP	设置调试程序，一般用于开发
SET_ORIENTATION	设置屏幕方向为横屏或以标准方式显示
SET_PREFERRED_APPLICATIONS	允许一个程序修改列表参数
SET_PROCESS_FOREGROUND	允许当前运行程序强行到前台
SET_PROCESS_LIMIT	允许程序设置最大的进程数量
SET_TIME	设置系统时间

（续）

权　限	说　明
SET_TIME_ZONE	设置系统时区
SET_WALLPAPER	允许程序设置壁纸
SET_WALLPAPER_HINTS	允许程序设置壁纸 hints
SIGNAL_PERSISTENT_PROCESSES	允许程序请求发送信号到所有显示的进程中
STATUS_BAR	允许程序打开、关闭或禁用状态栏及图标
SUBSCRIBED_FEEDS_READ	允许程序访问订阅的信息数据库
SUBSCRIBED_FEEDS_WRITE	写入或修改订阅内容的数据库
SYSTEM_ALERT_WINDOW	显示系统窗口
USE_CREDENTIALS	允许程序从 AccountManager 请求验证
USE_SIP	允许程序使用 SIP 视频服务
VIBRATE	允许振动
WAKE_LOCK	允许手机屏幕关闭后后台进程仍然运行
WRITE_APN_SETTINGS	写入网络 GPRS 接入点设置
WRITE_CALENDAR	写入日程，但不可读取
WRITE_CONTACTS	写入联系人，但不可读取
WRITE_SECURE_SETTINGS	允许程序读/写系统安全性的设置项
WRITE_OWNER_DATA	允许一个程序写入，但不读取所有者数据
WRITE_SETTINGS	允许程序读取或写入系统设置
WRITE_SMS	允许程序写短信
. WRITE_SYNC_SETTINGS	允许程序写入同步设置

【练习】

创建一个 Android 项目，要求如下：
1）项目名称使用自己姓名汉语拼音首字母的简写形式，如"张三"，即"zhs"；
2）修改 APP 应用的图标，并更改为圆形；
3）在界面上输出"安卓世界，我来啦！"；
4）对应用进行签名并打包为"welcome.jar"；
5）安装在自己的手机上并运行。

项目三　基础 UI 组件在 APP 界面中的运用

　　Android 提供了大量的、功能丰富的 UI 组件，开发者只要按照一定的规律把这些 UI 组件组合在一起，就可以开发出漂亮、优秀的图形用户界面（Graphics User Interface，GUI）。为了让这些 UI 组件能响应用户的诸如鼠标、键盘等动作，Android 也提供了类似 Java 的事件响应机制，这就可以保证图形界面应用可以响应用户的交互操作。

　　本项目在介绍 View 类的基础上，首先介绍 Android 界面编程的 3 种模式，即使用 XML 布局文件搭建 UI 界面、使用 Java 代码实现 UI 界面及使用 XML 和 Java 代码混合实现 UI 界面的方法；接着给出了自定义一个 View 的步骤；最后重点介绍了 Android 常用的基础 UI 组件的使用技巧和方法，如界面的布局方式、按钮、文本框等一些简单组件的使用。结合新闻 APP 中用户注册界面的实现，重点讲述使用网格布局管理器对界面进行布局的方法和步骤。

【知识目标】

- Android 中 View 类的作用
- 自定义 View 的方法和步骤
- Android 界面的编程模式
- Android 常用的几种布局管理器
- 线性、表格、网格、绝对布局管理器等的用法；
- TextView、EditView、CheckBox、RadidButton、Button 等 UI 组件的属性及用法

【模块分解】

- 使用 XML 和 Java 代码混合实现 UI 界面
- 继承 View 类实现一个自定义 View 组件——可以拖动的小球
- 使用 UI 布局管理器实现界面布局
- 使用文本框、按钮、复选框、单选按钮等实现用户注册界面

模块一　使用 XML 和 Java 代码混合实现 UI 界面

【模块描述】

　　Java 界面编程和 Android UI 编程在很多方面都很相似，可以相互借鉴经验，唯一不同的是，Android 应用是运行在手机等移动终端，Android UI 编程也需要调用 Android 的 API。可喜的是，Android 为开发者提供了强大的用户界面组件，借助这些组件，开发者可以很简单方便地开发出漂亮的用户界面。

知识点	技能点
➢ Android 的 UI 界面编程模式 ➢ View 类及子类的常用属性和方法 ➢ UI 界面设计的方法 ➢ ImageView 的简单用法 ➢ Android 事件监听机制	➢ 使用 XML 布局文件搭建 UI 界面 ➢ 使用 Java 代码实现 UI 界面 ➢ 使用 ImageView 显示图片 ➢ Android 事件监听器的添加方法 ➢ 使用 XML 布局文件和 Java 代码实现一个图片浏览器 UI 界面

任务 1　使用 XML 布局文件搭建 UI 界面

【任务描述】

用户可以通过使用 XML 布局文件来创建 UI 用户界面，此时只需要在 XML 文件中使用属性指定组件的属性，如 id 等。

本任务是创建一个 Android 应用，界面上显示一个"确定"按钮，现使用 XML 布局文件来实现本任务 UI 界面。

【任务实施】

1）创建一个 Android 工程 chap03_1，Application Name 为 TestViewActivity，如图 3-1 所示。

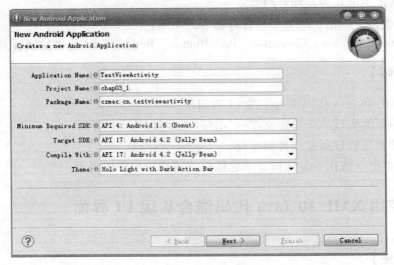

图 3-1　创建工程

默认生成的代码如下所示：

```
1.   public class TestViewActivity extends Activity {
2.       protected void onCreate(Bundle savedInstanceState) {
3.           super.onCreate(savedInstanceState);
```

```
4.          setContentView(R.layout.main);
5.     }
6. }
```

setContentView 方法用于指定一个布局文件，对 TestViewActivity 上的可视化组件进行创建和布局。布局文件都存放在 res/layout 文件夹下，可以使用"R.layout.文件名"的方式调用，如上述的 setContentView（R.layout.main），即使用 res/layout/main.xml 作为界面的布局文件。如果需要重新创建布局文件，可以通过修改 setContentView 方法的参数来调用。

2）在 res/layout 目录下创建 main.xml 文件，对界面进行布局。main.xml 文件的源文件如下所示：

```
1.  <RelativeLayout xmlns:android = "http://schemas.android.com/apk/res/android"
2.       xmlns:tools = "http://schemas.android.com/tools"
3.       android:layout_width = "match_parent"
4.       android:layout_height = "match_parent"
5.       android:paddingBottom = "@dimen/activity_vertical_margin"
6.       android:paddingLeft = "@dimen/activity_horizontal_margin"
7.       android:paddingRight = "@dimen/activity_horizontal_margin"
8.       android:paddingTop = "@dimen/activity_vertical_margin"
9.       tools:context = ".TestViewActivity" >
10.      <Button
11.          android:id = "@+id/btnOK"
12.          android:layout_width = "wrap_content"
13.          android:layout_height = "wrap_content"
14.          android:text = "确定" />
15. </RelativeLayout>
```

在这个源文件中，指定了布局方式为 LinearLayout 线性布局管理器（对线性布局管理器在后面章节将详细介绍）。在布局管理器中使用 <Button> 定义了一个按钮组件，指定按钮显示的文本内容为"确定"，同时指定了按钮的宽度和高度，并使用 android:id = "@+id" 指定了该按钮的 id 为"btnOK"。

3）运行 chap03_1，将显示如图 3-2 所示的界面。

main.xml 文件中定义了按钮的 id，这个 id 是该按钮的唯一标志，可以在其他资源文件或 Java 文件中使用该 id 来访问它，如在 Java 文件中访问指定的 UI 组件，可以通过如下代码实现：

```
findViewById(R.id.<android.id 属性值>);
```

一旦在程序中获取了指定的 UI 组件，接下来就可以通过代码来控制各 UI 组件了，包括外观定义，即绑定事件监听器等。

如在上述文件 TestViewActivity.java 中添加如下代码：

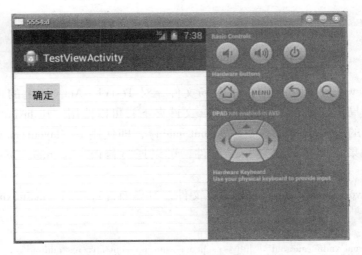

图 3-2 chap03_1 运行界面

```
1.   import android.view.View;
2.   import android.widget.Button;
3.   public class TestViewActivity extends Activity {
4.       Button btnOK;
5.       protected void onCreate(Bundle savedInstanceState) {
6.           super.onCreate(savedInstanceState);
7.           setContentView(R.layout.main);
8.           //获取界面中的 btnOK 按钮对象
9.           btnOK = (Button)this.findViewById(R.id.btnOK);
10.      }
11.      public void clickMe(View v)
12.      {
13.          btnOK.setText("我刚才被单击啦,哈");
14.      }
15.  }
```

如上述代码所示,第 1~2 行引入相关的包;第 4 行定义了一个按钮 btnOK;第 9 行使用 findViewById 方法获取界面中的按钮 btnOK,并将获取到的对象使用 Button 进行强制类型转换后将其赋予 btnOK;第 11~14 行定义了一个方法 clickMe,当调用这个方法时会将界面中的按钮的显示内容修改为"我刚才被单击啦,哈",但是如何做到当在界面中单击这个按钮时会调用这个方法呢? 具体步骤为如下。

将 main.xml 文件打开,可看到其中的 button 按钮增加了一个 onClick 属性,在这个属性中绑定这个方法即可,如下所示:

```
1.   <Button
2.       android:id = "@+id/btnOK"
3.       android:layout_width = "wrap_content"
```

```
4.          android:layout_height = "wrap_content"
5.          android:text = "确定"
6.          android:onClick = "clickMe"
7.     />
```

重新运行这个程序，将会出现如图 3-2 所示的运行界面，在界面中单击"确定"按钮，可以看到界面变成了如图 3-3 所示的内容。

图 3-3　chap03_1 运行界面

【知识学习】使用 XML 布局文件搭建 Android UI 界面时的注意事项

1) 布局文件只能是 XML 文件，不能是其他类型的，这个布局文件位于 res/layout 目录下。

2) Android 应用的 gen 目录下是系统自动生成的文件，其中包含一个类 R.java。在这个类中包括好多内部类，比如 layout 内部类，这个内部类里面定义的常量的名字与 res/layout 目录下的布局文件的名字相对应，如其中的 "public static final int main = 0x7f030000" 常量就对应 res/layout 目录下的布局文件 "main.xml"。

3) 在 Java 源码中可以使用 setContentView 方法，即通过 R.layout.main 指定 main.xml 布局文件。

4) 布局文件的根节点必须包含一个命名空间，且这个命名空间必须是 "xmls:android = "http://schemas.android.com/apk/res/android""。

5) 布局文件都可以通过 android:id 属性指定 id 值，指定 id 值的格式为 "@+id/value"。其中，"@+"表示如果 id 值在 R 类中的 id 内部类中不存在，则系统会自动产生这个 id 变量，如果在 R.id 类中存在这个变量，则可以直接使用这个变量。

任务 2　使用 Java 代码实现 UI 界面

【任务描述】

使用 XML 布局文件可以实现 UI 界面的布局，有"简单明了""将应用的视图控制器逻辑从Java业务逻辑代码中分离"及"更好地体现了MVC原则"等优点。但是如果开发者愿意,Andoird允许开发者像开发Java Swing桌面应用程序一样,完全使用Java代码实现UI界面。

【任务实施】

将上一任务修改,开发一个完全使用Java代码实现Android UI界面的应用。下面为其相应Activity代码内容:

```java
1.  public class TestViewActivityByJava extends Activity {
2.      protected void onCreate(Bundle savedInstanceState) {
3.          super.onCreate(savedInstanceState);
4.          //创建一个线性布局管理器
5.          LinearLayout layout = new LinearLayout(this);
6.          //设置该Activity显示这个layout
7.          super.setContentView(layout);
8.          layout.setOrientation(LinearLayout.VERTICAL);
9.          //创建一个按钮btnOK
10.         final Button btnOK = new Button(this);
11.         btnOK.setText("确定(chap03_2)");
12.         btnOK.setLayoutParams(new ViewGroup.LayoutParams(
13.                     ViewGroup.LayoutParams.WRAP_CONTENT,
14.                     ViewGroup.LayoutParams.WRAP_CONTENT));
15.         //向layout添加按钮btnOK
16.         layout.addView(btnOK);
17.         //为这个按钮绑定事件监听器
18.         btnOK.setOnClickListener(new OnClickListener()
19.         {
20.             public void onClick(View v)
21.             {
22.                 btnOK.setText("我刚才被单击了,哈,这个是使用Java代码实现的安卓UI界面");
23.             }
24.         }
25.         );
26.     }
27. }
```

上面的代码完全通过Java代码实现了Android界面,从中可以看出,每个视图组件都是一个View类型的对象,它们都可以在代码中使用视图类的构造方法,通过new关键字创建出来,并可以调用view对象的setXXX()方法设置其属性。

当运行chap03_2项目时,出现如图3-4所示的运行界面。

当单击上图中的"确定(chap03_2)"按钮时,将出现如图3-5所示的运行界面。

可见,要在Activity上显示并布局可视化组件,可以通过两种方式进行,一种是在res/

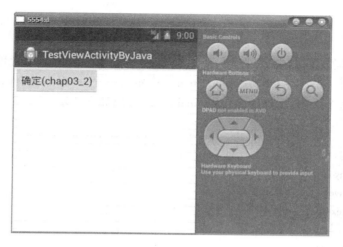

图 3-4　chap03_2 运行界面（a）

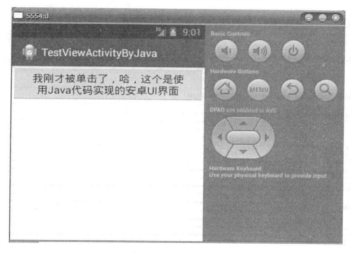

图 3-5　chap03_2 运行界面（b）

layout 目录下创建 XML 布局文件，另一种是使用 Java 代码在 Activity 中创建可视化组件，并设置给 Activity。

不难看出，使用第二种方式实现 UI 界面时，即使完全使用代码控制 UI 界面，也是不利于代码耦，而且由于通过 new 关键字创建 UI 可视化组件，需要调用 serXXX（）方法来设置 UI 组件的属性和行为，因此代码显得臃肿，编程比较烦琐。

任务 3　使用 XML 和 Java 代码实现图片浏览器

【任务描述】

通过前面的任务实现不难看出，如果在实现 Android UI 界面时完全使用 Java 代码，则编程烦琐且代码臃肿；如果完全使用 XML 布局文件来实现 Android UI 界面，虽然简单、便捷，但难免有失灵活。因此，很多情况下，使用 XML 布局文和 Java 代码混合的方式来实现 UI 界面。本任务是基于 XML 布局文件和 Java 代码实现一个简单的图片浏览器。

【任务实施】

1）在线性布局管理器中再创建一个线性布局管理器，这个布局管理器的 id 为 "second"。同样，在第一个线性布局管理器"root"中定义一个 textView 组件，用来显示"混合控制 UI 界面"，这个 XML 文件的 activity_main.xml 代码如下所示：

```
1.   <LinearLayout xmlns:android = "http://schemas.android.com/apk/res/android"
2.       xmlns:tools = "http://schemas.android.com/tools"
3.       android:layout_width = "fill_parent"
4.       android:layout_height = "fill_parent"
5.       tools:context = ".MainActivity"
6.       android:id = "@ + id/root"
7.       android:orientation = "vertical"
8.       >
9.
10.      <TextView
11.          android:id = "@ + id/title"
12.          android:layout_width = "wrap_content"
13.          android:layout_height = "wrap_content"
14.          android:text = "混合控制 UI 界面"/>
15.
16.      <LinearLayout
17.          android:id = "@ + id/second"
18.          android:layout_width = "fill_parent"
19.          android:layout_height = "fill_parent"
20.          >
21.      </LinearLayout>
22.  </LinearLayout>
```

2）编写 Activity 代码，即 MainActivity.java 代码如下：

```
1.   public class MainActivity extends Activity
2.   {
3.       //定义一个数组用来存放图片
4.       //it 为放在 drawable 目录下的图片名称，以下相同
5.       int[] images = new int[]{ R.drawable.it,
6.                                  R.drawable.jsp,
7.                                  R.drawable.web
8.                                };
9.       int currentImage = 0;
10.      @Override
11.      protected void onCreate(Bundle savedInstanceState)
12.      {
```

```
13.        super.onCreate(savedInstanceState);
14.        setContentView(R.layout.activity_main);
15.        //获取第二个布局管理器 second
16.        LinearLayout layout = (LinearLayout)findViewById(R.id.second);
17.        //创建一个 ImageView 组件,用来显示图片
18.        final ImageView imageView = new ImageView(this);
19.        //将图片组件放入 second 布局管理器中
20.        layout.addView(imageView);
21.        //初始化时显示第一张图片
22.        imageView.setImageResource(images[0]);
23.        imageView.setOnClickListener(new OnClickListener()
24.        {
25.            public void onClick(View v)
26.            {
27.                //当单击图片时切换图片
28.        imageView.setImageResource(images[++currentImage%images.length]);
29.            }
30.        });
31.    }
32. }
```

上面代码的 23～30 行表示为 ImageView 组件添加了一个单击事件监听器,当用户单击图片时将显示下一个图片,具体运行界面如图 3-6 所示。

图 3-6　chap03_3 运行界面

模块二 继承 View 类实现自定义 View 组件

【模块描述】

View 组件的作用类似于 Java 中的 AWT 组件或 Swing 组件中的面板 Panel 或 JPanel，是一个没有任何内容的一个矩形空白区域，每一个 View 都有一个用于绘图的画布，这个画布可以进行任意扩展。Android 中的其他组件都直接或间接地继承了 View 组件，然后在 View 组件上绘制外观。

知识点	技能点
➤ Android 的 UI 界面编程模式 ➤ View 类及子类的常用属性和方法 ➤ View 类中 onDraw 方法的作用 ➤ View 类中 onTouchEven 方法的使用 ➤ UI 界面设计的方法	➤ 继承 View 类实现自定义组件的方法和步骤 ➤ 自定义组件在 Android 界面中的使用方式

任务 自定义 View 组件实现可以随意拖动的小球

【任务描述】

在开发 Android 应用时，如果系统提供的 UI 组件不能很好地满足项目需求，可通过继承 View 类，自定义一个组件来满足项目需求。本任务是实现一个可以随手指移动的小球。

【任务实施】

1）创建项目，其参数设置如图 3-7 所示。

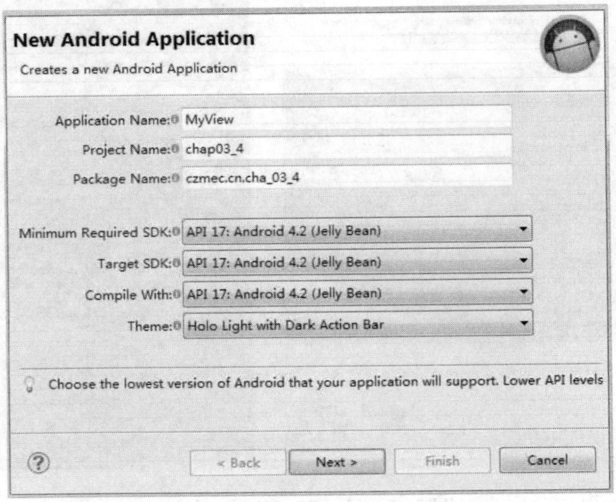

图 3-7 创建自定义的小球

2）单击 Next 按钮，直到进入如图 3-8 所示的界面。

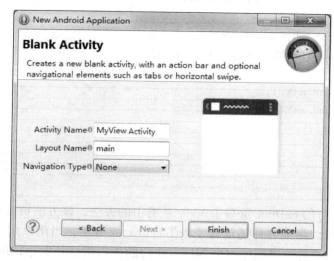

图 3-8　Activity 定义

3）在上图中输入 Activity 及 Layout 的名称，单击 Finish 按钮，项目创建完成。

4）在 src 文件夹下的"czmec. cn. cha_03_4"包中创建一个自定义组件类 MyView. java。此类继承了 View 类，并重写了其中的 onDraw 及 onTouchEvent 两个方法，还定义了一个构造方法，其具体代码如下所示：

```
1.   package czmec. cn. cha_03_4;
2.   public class MyView extends View
3.   {
4.       Paint mPaint;//画笔,包含了画几何图形、文本等的样式和颜色信息
5.       public float currentX = 40;
6.       public float currentY = 100;
7.       public MyView( Context context) {
8.           super( context);
9.           mPaint = new Paint( );
10.      }
11.
12.      public MyView( Context context, AttributeSet attrs) {
13.          super( context, attrs);
14.          mPaint = new Paint( );
15.      }
16.
17.      public void onDraw( Canvas canvas) {
18.          super. onDraw( canvas);
19.          mPaint. setStyle( Style. FILL);//设置填充
20.          mPaint. setColor( Color. RED);
```

```
21.              canvas.drawCircle(currentX, currentY, 15, mPaint);
22.          }
23.      //为该组件的触碰事件重写事件处理方法
24.      public boolean onTouchEvent(MotionEvent event)
25.      {
26.          currentX = event.getX();
27.          currentY = event.getY();
28.          this.invalidate();
29.              return true;
30.      }
31. }
```

上述代码第 17 行的 onDraw 重写了 View 基类中的方法,该方法负责在组件的指定位置绘制出一个红色小球。除此之外还重写了 onTouchEvent 方法,当用户用手或鼠标拖动该小球时,将会触发这个事件,进而使用第 24 ～ 30 行的代码重新获取小球的坐标值并使用 invalidate 方法通知当前组件重新绘制自己。

5)打开 XML 布局文件 main.xml,将自定义的 MyView 组件添加到线性布局管理器中,其代码如下所示:

```
1. <LinearLayout xmlns:android = "http://schemas.android.com/apk/res/android"
2.     android:orientation = "vertical"
3.     android:layout_width = "fill_parent"
4.     android:layout_height = "fill_parent"      >
5.     <czmec.cn.cha_03_4.MyView
6.         android:layout_width = "fill_parent"
7.         android:layout_height = "wrap_content" />
8. </LinearLayout>
```

6)运行 chap03_4,可以得到如图 3-9 所示的运行界面。

图 3-9 自定义小球运行界面

当使用鼠标或手指拖动小球时，发现小球可以随着拖动的轨迹运动，请试试看吧。

【知识学习】View 类

要设计一个用户界面，首先要熟悉界面上的组件，然后才可以将组件按照设计的要求进行合理的布局，最后要能够让组件响应用户的请求，即能够与用户交互。

读者已经知道，Android 应用中的每一个界面都是一个 Activity 类，而 Android 界面中展现的都是 Android 系统中的可视化组件，如按钮、文本框等组件，这些 Android 中的任何可视化组件都是 android.view.View 类的子类。图 3-10 所示为 Android API 文档的截图。

图 3-10　API 文档

可以看到，View 组件有很多直接或间接的子类，如 TextView、Button、CheckBox、ListView 等。这些子类都是 Android 应用中的常用视图组件，初学者需要了解 View 类中常用的方法及属性，并且要学会使用 API 帮助文档。

对于 View 类而言，它是所有可视化 UI 组件的基类，主要提供了控件绘制和事务处理的方法。创建用户界面所使用的控件都继承自 View，如 EditText、TextView、Button 等。它包含的 XML 属性和方法在所有组件中都可以使用。View 及其子类的相关属性，既可以在布局 XML 文件中进行相关设置，也可以通过成员方法在 Java 代码中动态设置。表 3-1 列出了 View 类常用的 XML 属性和方法。

表 3-1　View 类常用的 XML 属性及对应的方法

属性名称	对应方法	描述
android:background	setBackgroundResource(int)	设置背景颜色
android:clickable	setClickable(boolean)	设置 View 是否响应单击事件
android:visibility	setVisibility(int)	控制 View 的可见性

（续）

属性名称	对应方法	描述
android:fadingEdge	setVerticalFadingEdgeEnabled(boolean)	设置滚动该组件时组件边界是否使用淡出效果
android:fadingEdgeLength	getVerticalFadingEdgeLength()	设置淡出边界的长度
android:focusable	setFocusable(boolean)	控制View是否可以获取焦点
android:id	setId(int)	设置标识符可通过findViewById方法获取
android:longClickable	setLongClickable(boolean)	设置View是否响应长单击事件
android:layout_gravity		设置该组件在其容器中的对齐方式
android:layout_height	Set,LayoutParams(ViewGroup.LayoutParams p)	设置该组件在其父容器中的布局高度
android:layout_width	同上	设置该组件在其父容器中的布局宽度
android:onClick		为该组件的单击事件绑定监听器
android:soundEffectsEnabled	setSoundEffectsEnabled(boolean)	设置View触发单击等事件时是否播放音效
android:layout_margin		设置该组件在其父容器中布局时的页边距
android:minHeight		设置该组件的最小高度
android:minWidth		设置该组件的最小宽度
android:saveEnabled	setSaveEnabled(boolean)	如果未进行设置,当View被冻结时将不会保存其状态
android:nextFocusDown	setNextFocusDownId(int)	定义当向下搜索时应该获取焦点的View,如果该View不存在或不可见,则会抛出RuntimeException异常
android:nextFocusLeft	setNextFocusLeftId(int)	定义当向左搜索时应该获取焦点的View
android:nextFocusRight	setNextFocusRightId(int)	定义当向右搜索时应该获取焦点的View
android:nextFocusUp	setNextFocusUpId(int)	定义当向上搜索时应该获取焦点的View

View类有很多子类,可以分为3种,即布局类、视图容器类及视图类,其主要继承关系如图3-11所示。

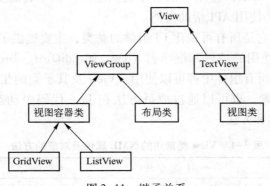

图3-11 继承关系

视图类指的是TextView、Button这样的类,它们不能再布局其他的组件,只能用来显示文字等。视图容器类都继承View的子类ViewGroup类,这些视图容器类也是一种视图类,但这些组件可以作为其他组件的容器，且 ViewGroup 又分为视图容器类和布局类。布局类用来管理组件的布局方式，如 LinearLayout、FrameLayout 等。这些内容在后续模块中将详细介绍。

【问题研讨】开发自定义组件的步骤

1）自己定义一个子类并继承 View 类。

2）重写构造方法。这是定制自己的 View 的最基本方式，当 Java 代码创建一个 View 实例对象时，需要调用这个构造方法。

3）重写 View 类中的一个或多个方法。

① onDraw（Canvas）：当该组件将要绘制它的内容时回调该方法进行绘制。
② onTouchEvent（MotionEvent）：当发生屏幕触摸事件时触发此方法。
③ onKeyDown（int，KeyEvent）：当某个键被按下的时候触发此方法。
④ onKeyUp（int，KeyEvent）：当松开某个键的时候触发此方法。
⑤ onSizeChanged（int，int，int，int）：当该组件的大小被改变时回调该方法。

上述的方法是常用的一些方法，当开发者需要自己定义 View 组件时，并不需要重写上述所有方法，只要根据业务的需要重写上述方法中的一个或几个方法即可。

4）如果自定义的 View 需要有自定义的属性，那么需要在 values 下建立 attrs.xml，在其中定义自己的属性。

5）使用自定义 View 的 XML 布局文件时需要加入"xmlns:前缀 = 'http://schemas.android.com/apk/res/自定义 View 所在的包路径'"。在使用自定义属性的时候使用"前缀:属性名"，如 my:textColor = "#FFFFFFFF"。

【任务拓展】

请读者认真思考下，可否通过在 Activity 中使用 Java 编码的方式将自定义小球添加到窗口中呢？如何修改上述代码实现？

模块三 使用 UI 布局管理器实现界面布局

【模块描述】

使用线性布局、表格布局、网格布局等管理器实现在 UI 界面布局。

知识点	技能点
➢ Android 线性、表格、网格等布局管理器的特点 ➢ 水平排列线性布局和垂直排列线性布局的区别 ➢ Android 布局管理器的分类 ➢ 线性、表格、网格等布局常用的 XML 属性和方法	➢ 使用水平或垂直线性布局管理器搭建界面布局 ➢ 在 Java 代码中通过 id 获取 XML 中组件的方法 ➢ 使用 Java 代码设置组件属性 ➢ 使用线性布局进行 UI 界面设计 ➢ 使用表格布局实现 UI 界面设计 ➢ 使用网格布局实现 UI 界面设计

任务1 使用线性布局实现在界面中动态添加按钮

【任务描述】

使用线性布局管理器 LinearLayout 实现在界面中动态添加若干按钮。

【任务实施】

1）创建一个 Android 应用，项目名称为"chap03_5"。打开"string.xml"文件，在 <resources> 节点中增加如下节点：

```
< string name = "button" > 按钮 </string >
```

2）打开 Activity，修改代码如下：

```
1.  public class MainActivity extends Activity {
2.      int count = 0;
3.      //计数器,记录按钮个数
4.      @Override
5.      public void onCreate(Bundle savedInstanceState) { //重写 onCreate 方法
6.          super.onCreate(savedInstanceState);
7.          setContentView(R.layout.activity_main);
8.          Button button = (Button) findViewById(R.id.btn);
9.          //获取屏幕中的按钮控件对象
10.         button.setOnClickListener(
11.         //为按钮添加 OnClickListener 接口实现
12.         new View.OnClickListener() {
13.             public void onClick(View v) {
14.                 LinearLayout ll = (LinearLayout) findViewById(R.id.lin);
15.                 //获取线性布局对象
16.                 String msg = MainActivity.this.getResources().getString(R.string.button);
17.                 Button tempbutton = new Button(MainActivity.this);
18.                 //创建一个 Button 对象
19.                 tempbutton.setText(msg + (++count)); // 设置 Button 控件显示的内容
20.                 tempbutton.setWidth(80);
21.                 //设置 Button 的宽度
22.                 ll.addView(tempbutton);
23.                 //向线性布局中添加 View
24.             }
25.         });
26.     }
27. }
```

3）打开布局文件 activity_main.xml，修改代码如下：

```
1.    < LinearLayout xmlns:android = "http://schemas.android.com/apk/res/android"
2.        android:orientation = "vertical"
3.        android:layout_width = "fill_parent"
4.        android:layout_height = "fill_parent"
5.        android:id = "@+id/lin"
6.        android:gravity = "right" >
7.
8.    < Button
9.        android:text = "添加"
10.       android:id = "@+id/btn"
11.       android:layout_width = "wrap_content"
12.       android:layout_height = "wrap_content"/ >
13.   </LinearLayout >
```

请读者注意上述 android:layout_width 及 android:layout_height 属性的用法。这两个属性分别指该组件的布局宽度和高度，它们支持如下 3 个属性值。

① fill_parent：指定子组件的高度或宽度及父容器组件的高度、宽度相同。

② match_parent：该属性值的含义与 fill_parent 完全相同，不过从 Android 2.2 开始不再推荐使用这个属性值，而采用 fill_parent 代替。

③ warp_content：指定子组件的大小恰好能包裹组件中的内容。

4）运行 chap03_5 项目，可以看到如图 3-12 所示的运行界面。

当连续 4 次单击上图的"添加"按钮时，可以看到如图 3-13 所示的运行界面。

图 3-12　chap03_5 运行界面（一）

图 3-13　chap03_5 运行界面（二）

5）将布局文件中的 android:orientation = "vertical"改为"horizontal"后，每单击一次"添加"按钮，就会在右方水平方向生成一个按钮。如连续单击 3 次后，运行的界面如图 3-14 所示。

6）当继续单击图 3-14 中的"添加"按钮时，出现了如图 3-15 所示的界面。

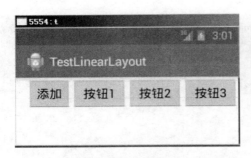
图 3-14　chap03_5 运行界面（三）

图 3-15　chap03_5 运行界面（四）

读者看到上图的按钮 3 右边的按钮 4 仅仅显示一点儿，这是什么原因呢？这是因为采用线性布局管理器时没有换行，当组件一个挨着一个排列到头之后，剩下的组件将会部分显示出来或完全显示不出来。

当该行在水平方向上容不下一个宽度为 80 的按钮时，按钮就会被压缩。此时再单击"添加"按钮时，画面没有任何变化，不会另起一行添加按钮，于是超出屏幕的将不会被显示。

【知识学习】UI 布局管理器及线性布局管理器

（1）UI 布局管理器

为了更好地管理 Android 应用的用户界面里的组件，Android 提供了布局管理器。通过布局管理器，Android 应用的图形用户界面具有良好的平台无关性。什么叫平台的无关性呢？就是说不同的手机它们的屏幕分辨率、尺寸并不完全相同，但使用 Android 的布局管理器来进行布局界面时它可以根据运行平台自动调整组件的大小，而开发者所需要做的就是选择合适的布局管理器。

布局管理器都是 ViewGroup 的子类，而 ViewGroup 也是 View 的子类，它们之间的关系如图 3-16 所示。

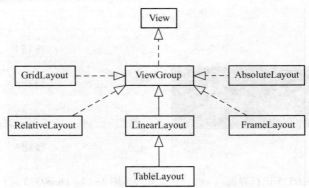

图 3-16　Android 布局管理器类图

从上图也可以发现，Android 4.2 布局管理器共有 6 种，即相对布局（RelativeLayout）、表格布局（TableLayout）、线性布局（LinearLayout）、帧布局（FrameLayout）、绝对布局（AbsoluteLayout）和网格布局（GridLayout）。所有布局都可作为容器类使用，因此可以调用多个重载的 addView 向布局管理器中添加组件，当然也可以用一个布局管理器嵌套其他布局管理器。

（2）线性布局管理器

线性布局（LinearLayout）管理器是非常常用的布局之一，它将放入其中的组件按照垂直或水平方向来布局，也就是控制放入其中的组件横向排列或纵向排列。在线性布局中，每一行（针对垂直排列）或每一列（针对水平排列）中只能放一个组件。并且Android的线性布局不会换行，当组件一个挨一个排列到窗体的边缘后，剩下的组件将不会被显示出来。

线性布局与AWT中FlowLayout的组件的区别在于：Android的线性布局不会换行，当组件一个挨着一个排到边缘时，剩下的组件将不会被显示出来；在AWT中，FlowLayout则会另起一行以便排列本行多出来的组件。

表3-2列出了LinearLayout的常用XML属性及相关方法。

表3-2 LinearLayout 常用 XML 属性及相关方法说明

XML 属性	相关方法	说 明
android:gravity	setGravity(int)	设置布局管理器内的组件对齐方式，该属性支持top、bottom、left、center_vertical等，可以同时指定多种对齐方式，多个属性值用竖线隔开，竖线前后不能有空格
android:orientation	setOrientation(int)	设置布局管理器内组件的排列方式，vertical 表示垂直，默认为 horizontal，表示水平
android:divider	setDividerDrawable(Drawable)	设置垂直布局时两个组件之间的分隔线

其中，Gravity 可取值的情况说明如表 3-3 所示。

表3-3 Gravity 取值说明

属 性 值	说 明
top	不改变空间大小，对齐到容器顶部
bottom	不改变空间大小，对齐到容器底部
left	不改变空间大小，对齐到容器左侧
right	不改变空间大小，对齐到容器右侧
center_vertical	不改变空间大小，对齐到容器纵向中央位置
center_horizontal	不改变空间大小，对齐到容器横向中央位置
center	不改变空间大小，对齐到容器中央位置
fill_vertical	若有可能，纵向拉伸以填满容器
fill_horizontal	若有可能，横向拉伸以填满容器
fill	若有可能，纵向与横向同时拉伸以填满容器

在使用 LinearLayout 线性布局管理器的时候，要注意的是，这个线性布局管理器中的所有子元素都受 LinearLayout.LayoutParams 控制，因此可以通过设定 LinearLayout 容器内包含的子元素的属性（如 layout_gravity）来设置子元素的外观或对齐方式。

任务 2 使用表格布局实现窗口布局

【任务描述】

基于表格布局实现界面布局，对表格中的数据填充采用两维数组实现。

【任务实施】

1）创建 chap03_6 项目，可使用 Java 编程的方式实现表格布局，具体为打开 Activity，修改代码如下：

```
1.  public class MainActivity extends Activity {
2.      private String titleData[][] = new String[][]{{"ID","姓名","邮箱","地址"},
3.          {"shl","孙华林","shl@czmec.com","江苏常州"},
4.          {"sy","孙岩","sy@czmec.com","江苏无锡"}};
5.      public void onCreate(Bundle savedInstanceState){
6.          super.onCreate(savedInstanceState);
7.          TableLayout tableLayout = new TableLayout(this);
8.          TableLayout.LayoutParams tableParams = new
9.  TableLayout.LayoutParams(ViewGroup.LayoutParams.MATCH_PARENT,
10.     ViewGroup.LayoutParams.MATCH_PARENT);
11.         tableLayout.setBackgroundResource(R.drawable.bg);
12.         //设置表格行
13.         for(int i=0;i<this.titleData.length;i++){
14.             TableRow tableRow = new TableRow(this);
15.             for(int j=0;j<this.titleData[i].length;j++){
16.                 TextView textView = new TextView(this);
17.                 textView.setText(this.titleData[i][j]);
18.                 tableRow.addView(textView,j);//加入一个编号
19.             }
20.             tableLayout.addView(tableRow);//向表格中增加若干表格行
21.         }
22.         super.setContentView(tableLayout,tableParams);
23.     }
24. }
```

第 11 行是为表格增加背景图片"bg.jpg"，第 13～21 行为设置表格行，第 20 行向表格中增加若干行。

2）运行 chap03_6 应用，可见如图 3-17 所示的界面。

【知识学习】表格布局管理器

表格布局（TableLayout）采用行列的形式管理 UI 组件，TableLayout 并不需要明确的声明包含多少行、多少列，而是通过添加 TableRow、其他组件来控制表格的行数和列数。

每次向 TableLayout 中添加一个 TableRow，该 TableRow 就是一个表格行。TableRow 也是容器，因此它也可以不断地添加其他组件，每添加一个子组件，该表格就增加一列。

如果直接向 TableLayout 中添加组件，那么这个组件将直接占用一行。

在表格布局中，列的宽度由该列中最宽的那个单元格决定，整个表格布局的宽度则取决与父容器的宽度（默认总是占满父容器本身）。

在表格布局管理器中，可以为单元格设置如下 3 种行为方式。

1) Shrinkable：如果某个列被设为 Shrinkable，那么该列的所有单元格的宽度可以被收缩，以保证表格能适应父容器的宽度。

2) Stretchable：如果某个列被设为 Stretchable，那么该列的所有单元格的宽度可以被拉伸，以保证组件能完全填满表格空余空间。

3) Collapsed：如果某个列被设为 Collapsed，那么该列的所有单元格会被隐藏。

TableLayout 继承了 LinearLayout，因此它完全可以支持 LinearLayout 所支持的全部 XML 属性，除此之外，TableLayout 还支持如表 3-4 所示的 XML 属性。

表 3-4　TableLayout 的常用 XML 属性及相关方法

XML 属性	相 关 方 法	说　　明
android:collapseColumns	setColumnCollapsed(int,boolean)	设置需要被隐藏的列的列序号，多个列序号之间用逗号隔开
android:shrinkColumns	setShrinkAllColumns(boolean)	设置允许被收缩的列的列序号，多个列序号之间用逗号隔开
android:stretchColumns	setStretchAllColumns(boolean)	设置允许被拉伸的列的列序号，多个列序号之间用逗号隔开

【任务拓展】

使用线性布局或表格布局来实现如图 3-18 所示的用户登录界面效果。

图 3-17　chap03_6 运行界面

图 3-18　用户登录界面

任务3 使用网格布局实现一个简易的计算器

【任务描述】

基于表格布局实现一个简易的计算器，要求界面中的部分按钮采用 Java 代码生成。

1. 任务实施

1）首先创建一个项目 chap03_7，Activity 的名称为 ComputerActivity，XML 布局文件的名称为 computer.xml。

2）设置窗口布局。

① 设置行列：设置 GridLayout 的 android:rowCount 为 6，设置 android:columnCount 为 4，这个网格为 6（行）* 4（列）的。

② 设置横跨四列：设置 TextView 和按钮横跨 4 列，即将 android:layout_columnSpan 的属性设置为 4，此时两个组件横跨表格 4 列，单独占了一行。

③ textView 的相关设置：设置 textView 中的文本与边框有 5 像素间隔，即 android:padding = "5px"。

具体的布局文件 computer.xml 的代码如下所示：

```
1.   <GridLayout xmlns:android = "http://schemas.android.com/apk/res/android"
2.       xmlns:tools = "http://schemas.android.com/tools"
3.       android:layout_width = "match_parent"
4.       android:layout_height = "match_parent"
5.       android:rowCount = "6"
6.       android:columnCount = "4"
7.       android:id = "@+id/root" >
8.       <!--
9.           定义一个6行*4列的GridLayout，在里面定义两个组件
10.          两个组件都横跨4列，单独占一行
11.      -->
12.      <TextView
13.          android:layout_width = "match_parent"
14.          android:layout_height = "wrap_content"
15.          android:layout_columnSpan = "4"
16.          android:textSize = "50sp"
17.          android:layout_marginLeft = "4px"
18.          android:layout_marginRight = "4px"
19.          android:padding = "5px"
20.          android:layout_gravity = "right"
21.          android:background = "#eee"
22.          android:textColor = "#000"
23.          android:text = "0"/>
24.      <Button
```

```
25.        android:layout_width = "match_parent"
26.        android:layout_height = "wrap_content"
27.        android:layout_columnSpan = "4"
28.        android:text = "清除"/>
29.
30.    </GridLayout>
```

3）打开 ComputerActivity 文件，编写 Activity 代码，思路如下。
① 指定组件所在行：GridLayout.SpecrowSpec = GridLayout.spec(int);
② 指定组件所在列：GridLayout.SpeccolumnSpec = GridLayout.spec(int);
③ 创建 LayoutParams 对象：
GridLayout.LayoutParams params = new GridLayout.LayoutParams(rowSpec,columnSpec);
④ 指定组件占满容器：params.setGravity(Gravity.FILL);
⑤ 将组件添加到布局中：gridLayout.addView(view,params);
其具体代码如下：

```
1.  package czmec.cn.computer;
2.  import android.os.Bundle;
3.  public class ComputerActivity extends Activity {
4.      private GridLayout gridLayout;
5.      //需要放到按钮上的字符串
6.      String chars[] = new String[]{
7.          "7", "8", "9", "/",
8.          "4", "5", "6", "*",
9.          "1", "2", "3", "-",
10.         ".", "0", "=", "+"
11.     };
12.     public void onCreate(Bundle savedInstanceState) {
13.         super.onCreate(savedInstanceState);
14.         setContentView(R.layout.computer);
15.         gridLayout = (GridLayout) findViewById(R.id.root);
16.         for(int i = 0; i < chars.length; i++) {
17.             Button button = new Button(this);
18.             button.setText(chars[i]);
19.             button.setTextSize(40);
20.             //指定组件所在行
21.             Spec rowSpec = GridLayout.spec(i/4 + 2);
22.             //指定组件所在列
23.             Spec columnSpec = GridLayout.spec(i % 4);
24.             //生成 LayoutParams 对象
25.             LayoutParams layoutParams = new LayoutParams(rowSpec, columnSpec);
26.             //指定组件充满网格
```

```
27.              layoutParams.setGravity(Gravity.FILL);
28.              //将组件设置于 GridLayout
29.              gridLayout.addView(button, layoutParams);
30.          }
31.      }
32. }
```

4)运行项目 chap03_7,结果如图 3-19 所示。

【知识学习】 网格布局管理器

在 Android 4.0 版本之前,如果想要达到网格布局(GridLayout)的效果,首先可以考虑使用最常见的 LinearLayout 布局,但是这样的排布会产生如下几点问题。

1)不能同时在 X、Y 轴方向上进行控件的对齐。

2)当多层布局嵌套时会对使用性能有影响。

3)不能稳定地支持一些可自由编辑布局的工具。

其次考虑使用表格布局 TabelLayout,这种方式会把包含的元素以行和列的形式进行排列,每行为一个 TableRow 对象,也可以是一个 View 对象。而在 TableRow 中还可以继续添加其他的控件,每添加一个子控件就成为一列。但是使用这种布局可能会出现不能使控件占据多个行或列的问题,而且渲染速度也不能得到很好的保证。

图 3-19 计算器界面

Android 4.0 以上版本出现的 GridLayout 布局管理器就解决了以上问题。GridLayout 将整个容器划分成 rows * columns 个网格布局,使用细虚线将布局划分为行、列和单元格,也支持一个控件在行、列上交错排列。而 GridLayout 使用的其实是跟 LinearLayout 类似的 API,只不过是修改了一下相关的标签而已,所以对于开发者来说,掌握 GridLayout 还是很容易的事情。

可 GridLayout 的布局策略简单地分为以下 3 个部分。

1)它与 LinearLayout 布局一样,也分为水平和垂直两种方式,默认是水平布局,一个控件挨着一个控件从左到右依次排列,但是通过指定 android:columnCount 设置列数的属性后,控件会自动换行进行排列。另一方面,对于 GridLayout 布局中的子控件,默认按照 wrap_content 的方式设置其显示,这只需要在 GridLayout 布局中显式声明即可。

2)若要指定某控件显示在固定的行或列,只需设置该子控件的 android:layout_row 和 android:layout_column 属性即可。但是需要注意:android:layout_row = "0" 表示从第一行开始,android:layout_column = "0" 表示从第一列开始,这与编程语言中一维数组的赋值情况类似。

3）如果需要设置某控件跨越多行或多列，只需将该子控件的 android:layout_rowSpan 或者 layout_columnSpan 属性设置为数值，再设置其 layout_gravity 属性为 fill 即可，前一个设置表明该控件跨越的行数或列数，后一个设置表明该控件填满所跨越的整行或整列。

表 3-5 所示为 GridLayout 的常用 XML 属性及相关方法。

表 3-5　GridLayout 的常用 XML 属性及相关方法

XML 属性	相关方法	说　　明
android:alignmentMode	setAlignmentMode(int)	设置网格布局管理器的对齐模式
android:columnCount	setColumnCount(int)	设置该网格布局的列数
android:columnOrderPreserved	setColumnOrderPreserved(boolean)	设置网格容器是否保留列序列号
android:rowCount	setRowCount(int)	设置该网格的行数
android:rowOrderPreserved	setRowOrderPreserved(int)	设置该网格容器是否保留行序列号
android:useDefaultMargins	setUseDefaultMargins(boolean)	设置该布局是否使用默认的页边距

为了控制该布局管理器中各子组件的布局分布，GridLayout 提供了一个内部类：GridLayout.LayoutParams，该类提供了大量的 XML 属性来控制 GridLayout 布局管理器中子组件的布局分布。表 3-6 所示为 GridLayout 参数表。

表 3-6　GridLayout 参数表

XML 属性	说　　明
android:layout_column	设置子组件在 GridLayout 中的列数
android:layout_columnSpan	设置该子组件在 GridLayout 中横向跨几列
android:layout_gravity	设置该组件采用何种方式占据该网格的空间
android:layout_row	设置该子组件在 GridLayout 中的行数
android:layout_rowSpan	设置该子组件在 GridLayout 中纵向横跨的行数

【任务拓展】

继续实现上述计算器的功能，使其实现加、减、乘、除的功能。

模块四　使用基础 UI 组件实现用户注册界面

【模块描述】

制作新闻 APP 的用户注册界面，它可以实现用户的注册功能，需要填写用户名、密码、密码确认、地址、E-mail、性别、爱好等用户信息。界面布局采用线性和表格嵌套布局。

知识点	技能点
➢ 线性布局和表格布局相关的 XML 属性和方法 ➢ 文本框 TextView 的属性和方法 ➢ 按钮 Button 的属性和用法 ➢ 复选框 CheckBox 的属性和方法 ➢ 单选按钮 RadioButton 的属性和用法	➢ 使用多种布局管理器搭建界面布局的技巧和方法 ➢ 文本框、按钮、复选框、单选按钮的用法 ➢ 界面设计/布局的技巧和方法

任务 用户注册界面的具体实现

【任务描述】

使用文本框、按钮、复选框、单选按钮、线性布局及表格布局管理器实现新闻 APP 用户注册界面。

【任务实施】

1）创建应用 chap03_8，设定 Activity 的名称为"UserRegisterActivity"，XML 布局文件的名称为"user_register"。其中，自动生成的布局文件界面如图 3-20 所示。

2）单击上图左边的"Layouts"选项，出现如图 3-21 所示的界面。

图 3-20 用户注册布局文件的可视化界面

图 3-21 Layouts 选项

3）在上图中选中"TableLayout"选项，将其拖动到图 3-20 的右边空白区域，将会在布局文件中增加一个表格布局管理器。用同样的方法，托运两个 TableRow 进入表格布局管理器，则分别增加两行。在所有行被选中的情况下，分别拖动文本框 TextView、按钮 Button 进入第一行和第二行，自动生成的布局文件代码如下所示：

```xml
1.  <TableLayout xmlns:android = "http://schemas.android.com/apk/res/android"
2.      android:layout_width = "fill_parent"
3.      android:layout_height = "fill_parent" >
4.      <TableRow
5.          android:id = "@+id/tableRow1"
6.          android:layout_width = "wrap_content"
7.          android:layout_height = "wrap_content" >
8.          <TextView
9.              android:id = "@+id/textView1"
10.             android:layout_width = "wrap_content"
11.             android:layout_height = "wrap_content"
12.             android:text = "TextView" />
13.         <EditText
14.             android:id = "@+id/editText1"
15.             android:layout_width = "wrap_content"
16.             android:layout_height = "wrap_content"
17.             android:ems = "10" >
18.             <requestFocus />
19.         </EditText>
20.     </TableRow>
21.     <TableRow
22.         android:id = "@+id/tableRow2"
23.         android:layout_width = "wrap_content"
24.         android:layout_height = "wrap_content" >
25.         <TextView
26.             android:id = "@+id/textView2"
27.             android:layout_width = "wrap_content"
28.             android:layout_height = "wrap_content"
29.             android:text = "TextView" />
30.         <EditText
31.             android:id = "@+id/editText2"
32.             android:layout_width = "wrap_content"
33.             android:layout_height = "wrap_content"
34.             android:ems = "10" />
35.     </TableRow>
36. </TableLayout>
```

对应的界面如图 3-22 所示。

4) 分别选中每个组件,在图 3-23 中分别更改每个组件对应的属性名。

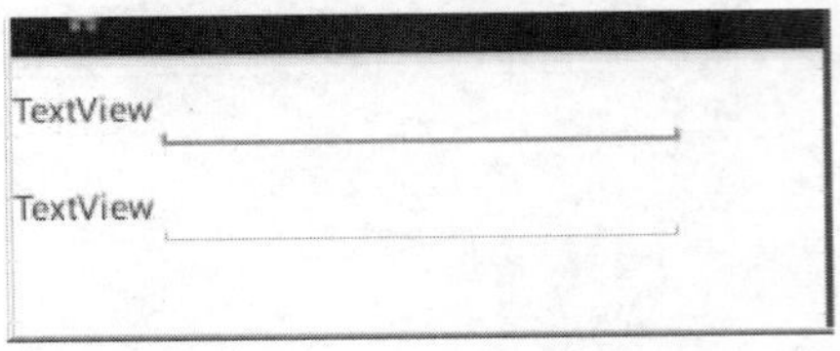

图 3-22　此时对应的用户注册界面

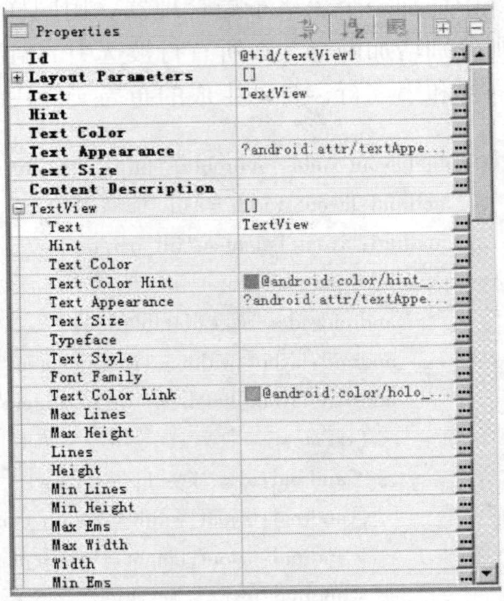

图 3-23　组件属性名修改

最后的布局文件程序代码如下所示:

```
1.   <TableLayout xmlns:android = "http://schemas.android.com/apk/res/android"
2.       android:layout_width = "fill_parent"
3.       android:layout_height = "fill_parent" >
4.    <TableRow
5.        android:id = "@+id/tableRow1"
6.        android:layout_width = "wrap_content"
7.        android:layout_height = "wrap_content" >
8.     <TextView
9.         android:id = "@+id/textView1"
10.        android:layout_width = "wrap_content"
11.        android:layout_height = "wrap_content"
12.        android:text = "用户名:" />
13.    <EditText
14.        android:id = "@+id/userName"
15.        android:layout_width = "wrap_content"
16.        android:layout_height = "wrap_content"
17.        android:ems = "10"
18.        android:hint = "请输入用户名" >
19.    </EditText>
20.    </TableRow>
21.    <TableRow
22.        android:id = "@+id/tableRow2"
```

```
23.          android:layout_width = "wrap_content"
24.          android:layout_height = "wrap_content" >
25.           < TextView
26.               android:id = "@+id/textView2"
27.               android:layout_width = "wrap_content"
28.               android:layout_height = "wrap_content"
29.               android:text = "用户密码:" />
30.           < EditText
31.               android:id = "@+id/userPassword"
32.               android:layout_width = "wrap_content"
33.               android:layout_height = "wrap_content"
34.               android:ems = "10"
35.               android:hint = "请输入密码"
36.               android:inputType = "textPassword" />
37.      </TableRow >
38.      < TableRow
39.          android:id = "@+id/tableRow3"
40.          android:layout_width = "wrap_content"
41.          android:layout_height = "wrap_content" >
42.           < TextView
43.               android:id = "@+id/textView3"
44.               android:layout_width = "wrap_content"
45.               android:layout_height = "wrap_content"
46.               android:text = "密码确认:" />
47.           < EditText
48.               android:id = "@+id/userPassword2"
49.               android:layout_width = "wrap_content"
50.               android:layout_height = "wrap_content"
51.               android:ems = "10"
52.               android:inputType = "textPassword" />
53.      </TableRow >
54.      < TableRow
55.          android:id = "@+id/tableRow4"
56.          android:layout_width = "wrap_content"
57.          android:layout_height = "wrap_content" >
58.           < TextView
59.               android:id = "@+id/textView4"
60.               android:layout_width = "wrap_content"
61.               android:layout_height = "wrap_content"
62.               android:text = "地址:" />
63.           < EditText
64.               android:id = "@+id/editText1"
```

```
65.            android:layout_width = "wrap_content"
66.            android:layout_height = "wrap_content"
67.            android:ems = "10"
68.            android:inputType = "textPostalAddress" />
69.     </TableRow>
70.     <TableRow
71.        android:id = "@+id/tableRow5"
72.        android:layout_width = "wrap_content"
73.        android:layout_height = "wrap_content" >
74.        <TextView
75.            android:id = "@+id/textView5"
76.            android:layout_width = "wrap_content"
77.            android:layout_height = "wrap_content"
78.            android:text = "E-mail:" />
79.        <EditText
80.            android:id = "@+id/editText2"
81.            android:layout_width = "wrap_content"
82.            android:layout_height = "wrap_content"
83.            android:ems = "10"
84.            android:inputType = "textEmailAddress" />
85.     </TableRow>
86.     <TableRow
87.        android:id = "@+id/tableRow6"
88.        android:layout_width = "wrap_content"
89.        android:layout_height = "wrap_content" >
90.        <TextView
91.            android:id = "@+id/textView6"
92.            android:layout_width = "wrap_content"
93.            android:layout_height = "wrap_content"
94.            android:text = "性别:" />
95.        <RadioGroup
96.            android:layout_width = "wrap_content"
97.            android:layout_height = "wrap_content"
98.            android:orientation = "horizontal" >
99.            <RadioButton
100.               android:id = "@+id/man"
101.               android:layout_width = "wrap_content"
102.               android:layout_height = "wrap_content"
103.               android:text = "男" />
104.           <RadioButton
105.               android:id = "@+id/woman"
106.               android:layout_width = "wrap_content"
```

```
107.            android:layout_height = " wrap_content"
108.            android:text = "女" />
109.        </RadioGroup >
110.      </TableRow >
111.      <TableRow
112.          android:id = " @ + id/tableRow7"
113.          android:layout_width = " wrap_content"
114.          android:layout_height = " wrap_content"
115.          >
116.          <TextView
117.              android:id = " @ + id/textView7"
118.              android:layout_width = " wrap_content"
119.              android:layout_height = " wrap_content"
120.              android:text = "爱好:" />
121.          <LinearLayout
122.              android:layout_width = " wrap_content"
123.              android:layout_height = " wrap_content" >
124.              <CheckBox
125.                  android:id = " @ + id/gj"
126.                  android:layout_width = " wrap_content"
127.                  android:layout_height = " wrap_content"
128.                  android:text = "逛街" />
129.              <CheckBox
130.                  android:id = " @ + id/kdy"
131.                  android:layout_width = " wrap_content"
132.                  android:layout_height = " wrap_content"
133.                  android:text = "看电影" />
134.          </LinearLayout >
135.      </TableRow >
136.      <TableRow
137.          android:id = " @ + id/tableRow8"
138.          android:layout_width = " wrap_content"
139.          android:layout_height = " wrap_content"
140.          android:gravity = " center" >
141.          <Button
142.              android:id = " @ + id/retister"
143.              android:layout_width = " wrap_content"
144.              android:layout_height = " wrap_content"
145.              android:gravity = " center"
146.              android:text = "注册" />
147.      </TableRow >
148. </TableLayout >
```

最后生成的运行界面如图 3-24 所示。

图 3-24　用户注册

【知识学习】TextView、EditView、Button 等组件

无论看上去多么美观的 UI 界面，开始都是先添加容器，然后不断地向容器中添加 UI 组件，最后形成一个美观的 UI 界面。Android 常用的 UI 组件有 TextView、EditView、Button、ImageButton、CheckBox、RadioButton 等。

TextView 组件和 EditView 组件是文本显示/编辑组件，前者只能用来显示文本，后者可以编辑文本。

（1）使用 TextView 显示文本

TextView 是最基本的用户界面元素，主要是用来在屏幕中显示固定长度的文本字符串内容，类似于 Swing 编程中的 Label 标签。

TextView 组件的类图关系如图 3-25 所示。

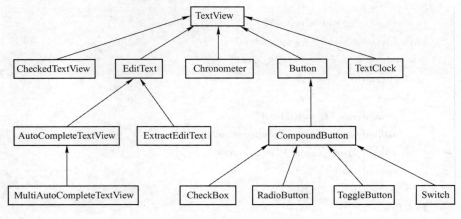

图 3-25　TextView 类图

从图 3-25 可见，TextView 还是 EditText、Button、CheckedTextView 等组件的父类。其中，CheckedTextView 组件增加了 checked 状态，可通过 setChecked（boolean）和 isChecked 方法访问及修改组件的 checked 状态，还可以通过调用 setCheckMarkDrawable 方法来设置它的勾选图标。

TextView 组件提供了大量的 XML 属性，这些属性不仅适用于 TextView，也适用于它的子类。这里仅仅介绍一些常用的属性。

1）android:autoLink：设置当文本为 URL 链接/E-mail/电话号码/map 时，文本是否显示为可单击的链接，其可选值为 none/web/email/phone/map/all。

2）android:autoText：设置在显示输入法并输入的时候，将自动执行输入值的拼写纠正。

3）android:capitalize：设置英文字母大写类型。需要弹出输入法才能看得到，参见 EditView 关于此属性的说明。

4）android:cursorVisible：设定光标为显示/隐藏，默认显示。

5）android:digits：设置允许输入哪些字符，如 "1、2、3、4、5、6、7、8、9、0、.、+、-、*、/、%、(、)"。

6）android:drawableBottom：在 text 的下方输出一个 drawable 资源。如果指定一个颜色的话，会把 text 的背景设为该颜色，并且同时和 background 使用时会覆盖后者。

7）android:drawableLeft：在 text 的左边输出一个 drawable 资源。

8）android:drawablePadding：设置 text 与 drawable（图片）的间隔，与 drawableLeft、drawableRight、drawableTop、drawableBottom 一起使用，可设置为负数，单独使用没有效果。

9）android:drawableRight：在 text 的右边输出一个 drawable。

10）android:drawableTop：在 text 的正上方输出一个 drawable。

11）android:editable：设置是否可编辑。

12）android:ellipsize：设置当文字过长时，该控件该如何显示参数 "start" 表示省略号显示在开头；"end" 表示省略号显示在结尾；"middle" 表示省略号显示在中间；"marquee" 表示以跑马灯的方式显示（动画横向移动）。

13）android:gravity：设置文本位置，如设置成 "center"，文本将居中显示。

14）android:hintText：为空时显示的文字提示信息，可通过 textColorHint 设置提示信息的颜色。

15）android:linksClickable：设置单击时是否链接，可设置为 autoLink。

16）android:marqueeRepeatLimit：在 ellipsize 指定 marquee 的情况下，设置重复滚动的次数，当设置为 marquee_forever 时表示重复滚动为无限次。

17）android:maxLength：限制显示的文本长度，超出部分不显示。

18）android:lines：设置文本的行数，设置两行就显示两行，即第二行没有数据。

19）android:maxLines：设置文本的最大显示行数，与 width 或者 layout_width 结合使用时表示超出部分自动换行，超出行数将不显示。

20）android:minLines：设置文本的最小行数，与 lines 类似。

21）android:password：以小点 "." 显示文本，值有 true 和 false。

22）android:phoneNumber：设置为电话号码的输入方式。

23）android:shadowColor：指定文本阴影的颜色，需要与 shadowRadius 一起使用。

24）android：shadowDx：设置阴影横向坐标开始位置。

25）android：shadowDy：设置阴影纵向坐标开始位置。

26）android：shadowRadius：设置阴影的半径。设置为 0.1 就变成字体的颜色了，一般设置为 3.0 的效果比较好。

27）android：singleLine：设置单行显示。如果与 layout_width 一起使用，表示当文本不能全部显示时，后面用"…"来表示。

28）android：text：设置显示文本。

29）android：textColor：设置文本颜色。

30）android：textColorHighlight：设置被选中文字的底色，默认为蓝色。

31）android：textColorHint：设置提示信息文字的颜色，默认为灰色。与 hint 一起使用。

32）android：textColorLink：设置文字链接的颜色。

33）android：textSize：设置文字大小，推荐度量单位为"sp"，如"15sp"。

34）android：textStyle：设置字形，包括 bold（粗体）、italic（斜体）、bolditalic（又粗又斜），可以设置一个或多个，用"|"隔开。

35）android：height：设置文本区域的高度，支持度量单位有 px（像素）/dp/sp/in/mm（毫米）。

36）android：width：设置文本区域的宽度，支持度量单位有 px（像素）/dp/sp/in/mm（毫米）。

37）android：maxWidth：设置文本区域的最大宽度。

38）android：minWidth：设置文本区域的最小宽度。

（2）使用 EditView 获取用户输入文本

EditText 与 TextView 具有很多相似的地方，它们最大的不同在于：TextView 不允许用户编辑文本内容，EditText 允许用户输入且可以编辑文本内容。

EditText 组件最重要的属性是 inputType，该属性相当于 HTML 的 <input…/> 元素的 type 属性。

EditText 派生如下两个子类。

1）AutoCompleteTextView：带有自动完成功能的 EditText，该类常与 Adapter 结合使用。

2）ExtractEditText：并不是 UI 组件，而是 EditText 组件的底层服务类，负责提供全屏输入法支持。

AutoCompleteTextView 将在讲解 Adapter 时详细介绍用法。

（3）按钮 Button

Android SDK 有两个简单的按钮控件：Button（android.widget.Button）和 ImageButton（android.widget.ImageButton）。这两个控件的功能很相似，因此可以一并地讨论它们。这两个控件不相的地方是外观上：Button 有一个文本标签，而 ImageButton 有一个可绘制的图像资源。

配置 Button 控件的外观时，需要在 Eclipse 中打开 XML 布局文件，首先选中该控件，此时在 Eclipse 的右下方的窗口会显示被选中组件的相关属性。此时，在属性标签中可以直接改变属性值来调节控件的属性，如图 3-26 所示。

图 3-26　改变组件属性

不仅仅是 Button 组件，Android 的所有其他的可视化组件都可以采用这种方法修改属性值。

下面是 Button 组件的一些重要的属性及设置注意事项。

1）使用 id 属性给 Button 或 ImageButton 一个唯一的名字。

2）使用文本属性设置 Button 控件上要显示的文字。

3）使用 src 属性设置 ImageButton 控件上要显示的图片。

4）将控件的布局高度和布局宽度属性设置为 wrap_content。

5）设置其他属性来调整控件的外观。比如，使用文本颜色、文本大小和文本样式属性来调整 Button 的字体。

（4）单选按钮及复选框

在 Android 中，单选按钮与复选框都是 Button 的子类，所以继承了 Button 的各种属性，而且还多了一个可选中的功能。

1）单选按钮 RadioButton。为什么叫单选按钮呢？因为只能选中一个，所以通常情况下，RadioButton 组件需要与 RadioGroup 组件一起使用，组成一个单选按钮组。RadioButton 的重要的属性为 checked，其值有 true 和 false 两种，表示单选按钮是否被选中。

具体使用的格式如下：

1. < RadioGroup
2. android:id = " @ + id/radioGroup1"
3. android:orientation = " horizontal"
4. android:layout_height = " wrap_connent"
5. android:layout_weithg = " wrap_connent" >

```
6.    <RadioButton
7.       android:id = "@ + id/ID 号"
8.       android:text = "显示的文本"
9.       android:checked = "true|false"
10.      android:layout_height = "wrap_connent"
11.      android:layout_weithg = "wrap_connent" >
12.   </RadioButton >
13.   …
14.   <!-- 可以添加多个 RadioButton -->
15. </RadioGroup >
```

2) CheckBox 复选框。相比于 RadioButton，复选框获取值就有点麻烦，因为用户可以多选，所以必须为每个复选框都设置一个 setOnCheckedChangeListener，具体用法将在后面详细介绍。

【任务拓展】

（1）帧布局

帧布局（FrameLayout）直接继承了 ViewCroup 组件。框布局容器为每个加入其中的组件创建一个空白的区域（称为一帧），每个子组件占据一帧，这些帧都会根据 gravity 属性执行自动对齐。也就是说，帧布局的效果有点类似于 AWT 编程的卡片布局 CardLayout，都是把组件一个一个地叠加在一起的。与 CardLayout 的区别在于，CardLayout 可以将下面的 Card 移上来，但 FrameLayout 则没有提供相应的方法。

表 3-7 所示为 FrameLayout 的常用 XML 属性及相关方法。

表 3-7 FrameLayout 的常用 XML 属性及相关方法说明

XML 属性	相 关 方 法	说　　明
android:foreground	setForeground(Drawable)	设置该框布局容器的前景图像
android:foregroundGravity	setForegroundGravity(int)	定义绘制前景图像的 gravity 属性
android:measureAllChildren	setMeasureAllChildren(boolean)	设置是否测量所有 children 的尺寸

（2）相对布局

相对布局（RelativeLayout）管理内子组件总是相对兄弟组件、它由父容器来决定的，因此这种布局方式被称为相对布局。

如果 A 组件的位置是由 B 组件的位置来决定的，那么 Android 要求先定义 B 组件，再定义 A 组件。表 3-8 所示为 RelativeLayout 的常用 XML 属性及相关方法。

表 3-8 RelativeLayout 的常用 XML 属性及相关方法说明

XML 属性	相 关 方 法	说　　明
android:gravity	setGravity(int)	设置该布局内部各子组件的对齐方式
android:ignoreGravity	setIgnoreGravity(int)	设置哪个组件不受 gravity 组件的影响

为了控制该布局管理器中各子组件的布局分布，RelativeLayout 提供了一个内部类，即 RelativeLayout.LayoutParams，该类提供了大量的 XML 属性来控制 RelativeLayout 布局管理器中子组件的布局分布。表 3-9 所示为 RelativeLayout 参数表。

表 3-9 RelativeLayout 参数表

XML 属性	说　　明
android:layout_above	将当前组件下边缘放置于参照组件之上，该属性为参照组件 ID
android:layout_alignBaseline	当前组件与参照组件的基线对齐，该属性为参照组件的 ID
android:layout_alignBottom	当前组件与参照组件的下边界对齐，该属性为参照组件的 ID
android:layout_alignLeft	当前组件与参照组件的左边界对齐，该属性为参照组件的 ID
android:layout_alignParenBottom	当前组件与父组件的下边界对齐，值为 true 或 false
android:layout_alignParentLeft	当前组件与父组件的左边界对齐，值为 true 或 false
android:layout_alignParentRight	当前组件与父组件的右边界对齐，值为 true 或 false
android:layout_alignParentTop	当前组件与父组件的上边界对齐，值为 true 或 false
android:layout_alignRight	当前组件与参照组件的右边界对齐，该属性为参照组件的 ID
android:layout_alignTop	当前组件与参照组件的上边界对齐，该属性为参照组件的 ID
android:layout_alignWithParentIfMissing	值为 true 或 false
android:layout_below	将当前组件上边缘放置于参照组件之下，该属性为参照组件 ID
android:layout_centerHorizontal	当前组件放置到父组件的水平居中的位置
android:layout_centerInParent	当前组件放置到父组件的重心位置
android:layout_centerVertical	当前组件放置到父组件垂直居中的位置
android:layout_toLeftOf	将当前组件右边缘放置于参照组件之下，该属性为参照组件 ID
android:layout_toRightOf	将当前组件左边缘放置于参照组件之下，该属性为参照组件 ID

（3）绝对布局

绝对布局（AbsoluteLayout）用于 Android 2.3.3 版本之前，现已废弃了。绝对布局就像 Java AWT 编程中的空布局，就是 Android 不提供任何布局控制，而是由开发人员自己通过 X 坐标、Y 坐标来控制组件的位置。当使用 AbsoluteLayout 作为布局容器时，布局容器不再管理子组件的位置、大小，这些都需要开发人员自己控制。

使用绝对布局时，每个子组件都可指定如下两个 XML 属性。

1）layout_x：指定该子组件的 X 坐标。

2）layout_y：指定该子组件的 Y 坐标。

android:layout_width 指定宽度是否充满父容器，或者恰恰正好包含子元素（子元素和容器大小相同）。android:width 是指组件的宽度，可以指定一个数字 + 单位的形式，如 100px 或者 100dp；同理，android:layout_height 和 android:height 类似。

Android 中有 px、dip、sp、in 等各种单位，这些单位介绍如下。

1）px：像素，每个 px 对应屏幕上的一个点，即实际的屏幕像素。

2）dip/dp：Device Independent Pixels，表示设备的独立像素，即相对于 160 dpi 屏幕的像素。这种单位是基于屏幕密度，在每英寸 160 点的显示器上，1 dp = 1 px，随着屏幕密度改变，dp 与 px 的换算会发生改变；

3）sp：Scale Pixels，比例像素，用于处理字体的大小，可以根据用户字体大小进行缩放。

4）in：英寸，标准长度单位。

5）mm：毫米，标准长度单位。

6）pt：磅，普通字体测量单位，标准长度单位，1/72 in。

当采用绝对布局来控制组件布局时，编程比较烦琐，且在不同的屏幕上的显示效果差异较大，因此建议读者尽量不要选用绝对布局进行界面设计。

【练习】

1. 基于 View 类，自定义开发的小球，实现一个自定义的正方形，要求如下：

1）使用 XML 和 Java 代码混合实现 Android UI 界面；

2）绘制一个正方形，设置其长和宽为 50 sp，并填充其中的颜色为蓝色；

3）在安卓虚拟手机中运行时，当按住鼠标左键拖动时，正方形可以随着鼠标拖动轨迹移动；如果安装在 Android 真机中，则正方形可以随着手指在屏幕上滑动。

2. 使用线性布局、相对布局、表格布局等布局管理器，结合 EditText、ImageView、TextView、ImageButton、CheckBox、Button 等 Android 组件，实现一个 QQ 登录布局界面，效果图如图 3-27 所示。

3. 使用框布局实现如图 3-28 所示的界面。

图 3-27　QQ 登录界面

图 3-28　框布局实现的界面

4. 在 Layout 布局内添加一个长和宽为 200 dp 的红色正方形，效果图如图 3-29 所示。

5. 在布局中加入 radioGroup 控件（包含 3 个垂直排列的 radioButton，名字为 radio1、radio2、radio3）和 textView 控件，当选中其中一个 radioButton 后，textView 内会显示被选中的 radioButton 名称，如图 3-30 所示。

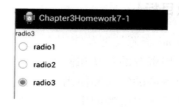

图 3-29　正方形预期效果图　　　　图 3-30　radioButton 的使用

项目四　各种 Android 资源在 APP 应用中的引用

资源是 Android 应用中的重要组成部分。Android 应用程序可以使用各种资源设置颜色、字体大小、风格等属性，也可以是资源存储图像、数组等。本项目将详细讲解 Android 中的各种资源及使用。

【知识目标】

- Android 资源类型及使用方式
- 字符串资源的定义和使用
- 颜色资源的定义和使用
- 尺寸、数组资源的定义和使用
- Drawable 资源的定义和使用
- Android 系统资源及 assets 资源的定义和使用

【模块分解】

- 字符串/颜色/尺寸资源在 UI 界面中的使用
- 数组资源在 UI 界面中的使用
- Drawable 资源在 UI 界面中的使用
- assets 及 Android 系统资源在 UI 界面中的使用
- 基于资源引用方式优化新闻 APP 用户注册页面

模块一　字符串/颜色/数组等基础资源的定义和使用

【模块描述】

字符串、颜色、尺寸等资源属于简单资源类型，它们对应的 XML 文件都位于/res/values 目录下，其默认的文件名分别为 strings、colors、dimens。当然，如果开发人员愿意，可以将相同类型的资源如 colors.xml 文件创建成 bright_colors.xml 和 muted_colors.xml 两个 XML 文件来存储颜色。数组资源文件位于 res\values 目录下，一般用 arrays.xml 文件名来定义数组。Drawable 资源是保存在 res/drawable 目录下的，该资源是 Android 应用中使用最频繁的资源。用户不仅可以直接使用 png、jpg、gif、jpeg 等图片作为资源，也可以使用多种 XML 文件作为资源，只要将 XML 文件编译成 Drawable 子类的对象即可作为 Drawable 资源使用。本项目主要讲述这些资源的定义和使用。

知识点	技能点
➤ Android 资源分类及颜色值 ➤ 颜色资源、字符串资源、尺寸资源 ➤ 资源的定义和使用分离的好处 ➤ 图片资源及 Drawable 资源的使用 ➤ 布局资源的定义和使用	➤ 颜色资源、字符串资源、尺寸资源的定义和使用方法 ➤ 使用 Eclipse ADT 可视化设置简单资源的方法 ➤ 使用 Eclipse ADT 可视化引用简单资源的方法 ➤ 在 Java 代码中获取及使用图像资源的方法

任务 1　基于资源引用方式优化用户注册页面

【任务描述】

将 APP 界面中的字符串常量、颜色、尺寸等放入相应的 XML 中，然后在界面中引用，显示与数据分离，优化了界面。

开发人员可以通过手工编写资源文件的 XML 文件，诸如字符串、颜色及尺寸等简单资源后，再使用 AAPT 工具编译它们，并生成 R.java 文件。当然，也可以通过 ADT 插件更简单地实现。下面将详细介绍使用 Eclipse 内置的 ADT 插件设置字符串、颜色及尺寸资源文件的方法和步骤，优化 chap03_8 应用（模块三中的 APP 用户注册界面）。

【任务实施】

1）将 chap03_8 工程项目导入到 Eclipse 中，定位/res/values/strings.xml 文件，双击打开并编辑它，strings.xml 资源文件将在右边的面板中显示，如图 4-1 所示。

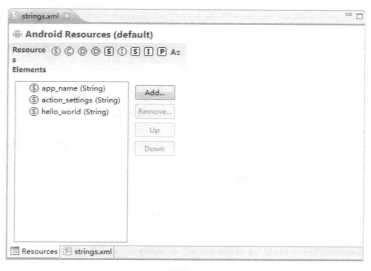

图 4-1　Eclipse 资源编辑器显示的资源文件

上图的底部有两个选项卡，即"Resources"和"strings.xml"，前者提供了可视化地编辑 XML 文件的方法，如可以插入字符串、颜色及尺寸资源，具体操作方法为单击上图中的 Add、Remove 等按钮进行 XML 节点的添加和删除。如单击 Add 按钮，可弹出如图 4-2 所示的界面，在这个界面中可以添加 XML 节点并指定其值。

当选中 String 选项后，则会在 strings.xml 文件中添加一个 <string/> 节点，然后得到图 4-3 所示的界面，输入节点的 name 和 value 即可。

图 4-2　可视化添加 XML 节点

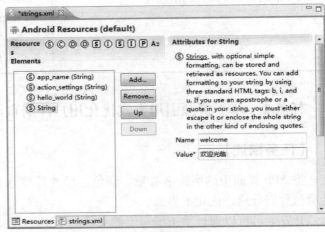
图 4-3　设置 XML 节点信息

打开"strings.xml"选项卡，可以显示创建的 XML 资源文件的源码，并直接手动编辑 XML 资源文件，如图 4-4 所示。

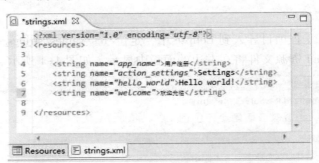

图 4-4　strings.xml 资源文件源码

2）按照上述方法及步骤，打开 strings.xml 资源文件，修改的内容如下：

1. <?xml version = "1.0" encoding = "utf - 8"?>
2. <resources>
3. 　　<string name = "app_name">用户注册</string>
4. 　　<string name = "action_settings">Settings</string>
5. 　　<string name = "hello_world">Hello world!</string>
6. 　　<string name = "welcome">欢迎光临</string>

```
7.      < string name = "userNameText" >用户名：</string >
8.      < string name = "passwordText" >用户密码：</string >
9.      < string name = "userNameHintText" >请输入用户名：</string >
10.     < string name = "passwordHintText" >请输入密码：</string >
11.     < string name = "password2Text" >密码确认：</string >
12.     < string name = "addressText" >地址：</string >
13.     < string name = "emailText" >E – mail：</string >
14.     < string name = "sexText" >性别：</string >
15.     < string name = "man" >男</string >
16.     < string name = "woman" >女</string >
17.     < string name = "aihao" >爱好：</string >
18.     < string name = "guangjie" >逛街</string >
19.     < string name = "kandianying" >看电影</string >
20.     < string name = "reg" >注册</string >
21. </resources >
```

以上程序表示将所有 Android 界面中显示的字符串内容都移到这个 strings.xml 资源文件。

3）在/res/values/目录下创建一个 colors.xml 资源文件，用于存放相关颜色资源。具体为打开 strings.xml 资源文件，修改的内容如下：

```
1.  < ?xml version = "1.0" encoding = "utf – 8"? >
2.  < resources >
3.      < color name = "regButtonFontColor" >#ff0000 </color >
4.      < color name = "hintFontColor" >#ff0000 </color >
5.  </resources >
```

4）打开/res/values/目录下的 dimens.xml 资源文件，这个资源文件用来存放尺寸资源。具体为打开 strings.xml 资源文件，增加 regButtonSize 变量，用来设置注册按钮中字体的大小，修改的内容如下。

```
1.  < resources >
2.      < dimen name = "activity_horizontal_margin" >16dp </dimen >
3.      < dimen name = "activity_vertical_margin" >16dp </dimen >
4.      < dimen name = "regButtonSize" >30dp </dimen >
5.  </resources >
```

5）将"btnbackground.png"图片文件复制到/res/drawable – mdpi/目录下，用于注册按钮的背景图片。

6）在 XML 资源文件中引用上述定义的字符串、颜色、尺寸、图片等相关资源。打开/res/layout/目录下的 userregister.xml 布局文件，在"GraphiclaLayout"选项卡下，采用可视化图形方式修改相关资源引用。下面以引用图片及颜色资源为例进行操作说明。

① 在 userregister.xml 布局文件的"GraphiclaLayout"选项卡中，首先用鼠标选中 reg 按

钮，此时，这个按钮的相关属性会显示在右下角的 Properties 面板中，如图 4-5 所示。

② 在这个窗口中单击 按钮，对属性进行顺序排列，然后找到 BackGround、Text、TextColor、TextSize 属性，分别对按钮的背景图片、显示文字、字体颜色及字体大小进行资源引用。如单击 BackGround 右边的 按钮，弹出如图 4-6 所示的对话框。

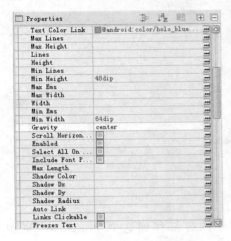

图 4-5　按钮属性面板

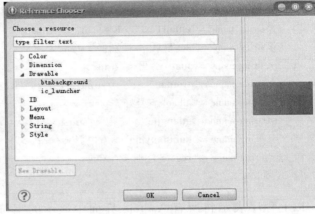

图 4-6　图片选择对话框

③ 选中 drawable 目录下的 "btnbackground"，就可以对按钮的背景进行设置了。同样，其他的资源设置方法类同。

7）通过上述的可视化资源设置，会自动生成如下 layout 布局文件内容：

```
1.    <TableLayout xmlns:android = "http://schemas.android.com/apk/res/android"
2.        android:layout_width = "fill_parent"
3.        android:layout_height = "fill_parent" >
4.        <TableRow
5.            android:id = "@+id/tableRow1"
6.            android:layout_width = "wrap_content"
7.            android:layout_height = "wrap_content" >
8.            <TextView
9.                android:id = "@+id/textView1"
10.               android:layout_width = "wrap_content"
11.               android:layout_height = "wrap_content"
12.               android:text = "@string/userNameText" />
13.           <EditText
14.               android:id = "@+id/userName"
15.               android:layout_width = "wrap_content"
16.               android:layout_height = "wrap_content"
17.               android:ems = "10"
18.               android:hint = "@string/userNameHintText"
19.               android:textColorHint = "@color/hintFontColor" >
20.           </EditText>
```

```
21.        </TableRow>
22.        <TableRow
23.            android:id = "@+id/tableRow2"
24.            android:layout_width = "wrap_content"
25.            android:layout_height = "wrap_content" >
26.            <TextView
27.                android:id = "@+id/textView2"
28.                android:layout_width = "wrap_content"
29.                android:layout_height = "wrap_content"
30.                android:text = "@string/passwordText" />
31.            <EditText
32.                android:id = "@+id/userPassword"
33.                android:layout_width = "wrap_content"
34.                android:layout_height = "wrap_content"
35.                android:ems = "10"
36.                android:hint = "@string/passwordHintText"
37.                android:textColorHint = "@color/hintFontColor"
38.                android:inputType = "textPassword" />
39.        </TableRow>
40.        <TableRow
41.            android:id = "@+id/tableRow3"
42.            android:layout_width = "wrap_content"
43.            android:layout_height = "wrap_content" >
44.            <TextView
45.                android:id = "@+id/textView3"
46.                android:layout_width = "wrap_content"
47.                android:layout_height = "wrap_content"
48.                android:text = "@string/password2Text" />
49.            <EditText
50.                android:id = "@+id/userPassword2"
51.                android:layout_width = "wrap_content"
52.                android:layout_height = "wrap_content"
53.                android:ems = "10"
54.                android:inputType = "textPassword" />
55.        </TableRow>
56.        <TableRow
57.            android:id = "@+id/tableRow4"
58.            android:layout_width = "wrap_content"
59.            android:layout_height = "wrap_content" >
60.            <TextView
61.                android:id = "@+id/textView4"
62.                android:layout_width = "wrap_content"
```

```
63.            android:layout_height = "wrap_content"
64.            android:text = "@string/addressText" />
65.        <EditText
66.            android:id = "@+id/address"
67.            android:layout_width = "wrap_content"
68.            android:layout_height = "wrap_content"
69.            android:ems = "10"
70.            android:inputType = "textPostalAddress" />
71.    </TableRow>
72.    <TableRow
73.        android:id = "@+id/tableRow5"
74.        android:layout_width = "wrap_content"
75.        android:layout_height = "wrap_content" >
76.        <TextView
77.            android:id = "@+id/textView5"
78.            android:layout_width = "wrap_content"
79.            android:layout_height = "wrap_content"
80.            android:text = "@string/emailText" />
81.        <EditText
82.            android:id = "@+id/email"
83.            android:layout_width = "wrap_content"
84.            android:layout_height = "wrap_content"
85.            android:ems = "10"
86.            android:inputType = "textEmailAddress" />
87.    </TableRow>
88.    <TableRow
89.        android:id = "@+id/tableRow6"
90.        android:layout_width = "wrap_content"
91.        android:layout_height = "wrap_content" >
92.        <TextView
93.            android:id = "@+id/textView6"
94.            android:layout_width = "wrap_content"
95.            android:layout_height = "wrap_content"
96.            android:text = "@string/sexText" />
97.        <RadioGroup
98.            android:layout_width = "wrap_content"
99.            android:layout_height = "wrap_content"
100.           android:orientation = "horizontal" >
101.           <RadioButton
102.               android:id = "@+id/man"
103.               android:layout_width = "wrap_content"
104.               android:layout_height = "wrap_content"
```

```
105.            android:text = "@string/man" / >
106.        < RadioButton
107.            android:id = " @ + id/woman"
108.            android:layout_width = " wrap_content"
109.            android:layout_height = " wrap_content"
110.            android:text = "@string/woman" / >
111.      < /RadioGroup >
112.    < /TableRow >
113.    < TableRow
114.        android:id = " @ + id/tableRow7"
115.        android:layout_width = " wrap_content"
116.        android:layout_height = " wrap_content" >
117.        < TextView
118.            android:id = " @ + id/textView7"
119.            android:layout_width = " wrap_content"
120.            android:layout_height = " wrap_content"
121.            android:text = "@string/aihao" / >
122.        < LinearLayout
123.        android:layout_width = " wrap_content"
124.        android:layout_height = " wrap_content" >
125.        < CheckBox
126.            android:id = " @ + id/gj"
127.            android:layout_width = " wrap_content"
128.            android:layout_height = " wrap_content"
129.            android:text = "@string/guangjie" / >
130.        < CheckBox
131.            android:id = " @ + id/kdy"
132.            android:layout_width = " wrap_content"
133.            android:layout_height = " wrap_content"
134.            android:text = "@string/kandianying" / >
135.        < /LinearLayout >
136.    < /TableRow >
137.    < TableRow
138.        android:id = " @ + id/tableRow8"
139.        android:layout_width = " wrap_content"
140.        android:layout_height = " wrap_content"
141.        android:gravity = " center" >
142.        < Button
143.            android:id = " @ + id/retister"
144.            android:layout_width = " wrap_content"
145.            android:layout_height = " wrap_content"
146.            android:background = "@drawable/btnbackground"
```

```
147.                android:gravity = "center"
148.                **android:text = "@string/reg"**
149.                **android:textColor = "@color/regButtonFontColor"**
150.                **android:textSize = "@dimen/regButtonSize" / >**
151.            </TableRow >
152.        </TableLayout >
```

上述加粗的程序代码表示在 XML 文件中对资源的引用,可以用 ADT 开发平台的可视化操作自动生成这部分代码。当然,也可以打开文本编辑器直接修改。此时,运行项目,可看到如图 4-7 所示的运行界面。

【知识学习】 资源类型及使用

(1) Android 应用中的文件分类

Android 应用中的文件大致分为以下 3 种。

1) 布局文件:即界面布局文件,这是一个 XML 文件,主要用来对 Android 的界面进行布局。布局文件中每个节点都对应相应的 View 标签。

2) Java 源代码文件:一般对于 Android 初级程序员来说,主要有两种源代码文件,即 Activity、封装业务逻辑或封装数据的 JavaBean。

3) 资源文件:Android 中存在大量的资源,很多资源都会被封装到 apk 文件中,并随 apk 文件一起发布。这些资源不仅包括各种类型的图片资源,更重要的是还有以各种 XML 为主的资源,如字符串资源、颜色资源、数组资源、菜单资源等。这些资源一般都存放在 Android 应用的 res 目录和 assets 目录。存放在 res 目录下的资源会在编译时自动在 R.java 文件中为这些资源创建索引,这样,在程序中可以直接通过 R 资源清单类进行访问;而存放在 assets 目录下的资源则是 Android 应用无法直接访问的原生资源,需要通过 AssetManager 类以二进制流的形式来读取。

在前面实现的案例中,读者可以发现,很多情况下是在 XML 布局文件或者 Java 代码中直接使用一些字符串等常量,这实际上并不是一种好的编程方式。编写良好的代码往往是在程序中调用资源文件,而不是直接写在代码中,这样做使得代码更容易阅读及管理。接下来的内容将为读者详细介绍 Android 资源文件的使用,进而改进、优化前面的编程方式。

(2) 资源类型

资源是 Android 应用的重要组成部分,字符串、图像、声音、视频等都是资源。除了图形、视频等使用原始文件作为资源存储外,大多数常见的资源类型都存放在 XML 文件中,这些资源都保存在 assets 目录及 res 目录的子目录下。

1) assets 目录下主要存放的是原生资源,可以是任意类型的资源,这些资源不会被编译,Android 应用也无法直接访问,可以通过 AssetManager 类以二进制流的形式来读取。

2) res 目录下的资源会在编译时自动在 R.java 文件中为这些资源创建索引,这样在程序中可以直接通过 R 资源清单类进行访问。

res 目录有几个子目录,即 layout 目录、values 目录、drawable 目录及 menu 目录。这些目录分别用来存储不同类型的资源。

通过前面的介绍，读者已经了解了 Android 应用程序的结构，在这个结构中可以看到资源分布及存储情况，如图 4-8 所示。

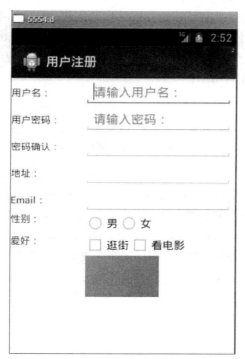

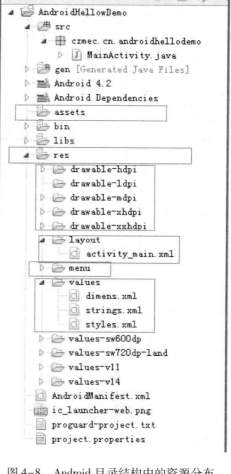

图 4-7　运行界面　　　　　　　　图 4-8　Android 目录结构中的资源分布

前面创建的 Android 应用是由基于 ADT 插件的 Eclipse 工具创建的，读者会发现在应用中加入资源是非常简单的操作。当开发者将 Android 应用的 res 目录下的某个文件添加新的资源时，ADT 插件会自动探知并在后台编译这些新的资源，并在 R.java 文件中生成对应的常量，开发者在实现功能时就可以直接使用这些资源常量了。

下面对上图中生成的资源目录分布进行介绍。

① layout 子目录用来存放资源类型为 XML 的布局文件，如图中的"activity_main.xml"。布局文件的名字可以随便取，但是必须是 XML 类型的文件。这个目录下的所有布局文件在编译时都会在 R.java 类文件中生成一个以此布局文件名字命名的常量。如"activity_main.xml"文件在 R.java 文件中生成对应的常量，且这个常量是在以一个名字为"layout"命名的内部类中定义的，如下代码所示：

```
1.  public static final class layout
2.  {
3.          public static final int activity_main = 0x7f030000;
4.  }
```

② res 目录下的 drawable 子目录用来存放资源类型为图片的资源或各种 Drawable 对象的 XML 文件。图片可以是任何格式的文件，如扩展名为 .png、.jpg、.jpeg、.gif 等格式的文件。读者会发现，在图 4-8 中包括了 drawable – ldpi、drawable – mdpi、drawable – hdpi、drawable – xhdpi 及 xxhdpi 这 5 种目录。这是在 Android 4.2 版本及更高的版本出现的一种现象，即分别提供了低分辨率、中分辨率、高分辨率、超高分辨率等目录。这些目录中可以存放相同名称而分辨率不同的图片，系统在运行时会根据屏幕的分辨率自动加载相应文件夹下的图片。除此之外，drawable 目录下还可以存放各种 Drawable 对象的 XML 文件，如 BitmapDrawable。不管是开发者自定义的 Drawable 对象的 XML 文件，还是图片文件，在系统编译时都会在 R.java 类文件生成对应的常量，如下代码所示：

```
1.  public static final class drawable
2.  {
3.          public static final int ic_launcher = 0x7f020000;//ic_launcher 为图片的名称
4.  }
```

③ values 子目录用来存放资源类型为 XML 的各种 XML 文件。这些文件用来存放各种简单的常量，如字符串常量、颜色值、数组、尺寸、样式等。如图 4-8 所示，Android 4.2 为项目自动生成了 3 种类型的 XML 文件，即 strings.xml、styles.xml 和 dimens.xml。这些文件分别用来存放不同的类型的常量。下面罗列出了各种常用的 XML 文件存放的值及类型。

- arrays.xml：定义数组资源；
- colors.xml：定义颜色常量资源；
- dimens.xml：定义尺寸资源；
- strings.xml：定义字符串常量资源；
- styles.xml：定义样式资源。

这些资源文件中定义的常量也会在 R.java 类中自动生成，如在 string.xml 文件中定义的常量会自动生成如下代码：

```
1.  public static final class string
2.  {
3.          public static final int action_settings = 0x7f050001;
4.          public static final int app_name = 0x7f050000;
5.          public static final int hello_world = 0x7f050002;
6.  }
```

④ menu 子目录用来存放各种菜单资源，如选项菜单、子菜单、上下文菜单等。

⑤ 除此之外，还可以在 res 目录中自己定义一个 xml 子目录，用来存放任意的原生文件，这些 XML 文件可以在 Java 代码中使用 Resources.getXML 方法进行访问。

需要开发者注意的是，上述 res 目录下的资源文件名称的定义有严格的限制，比如只能由字母、数字和下画线组成且字母必须小写。

（3）资源的使用

在 Android 应用中，资源使用非常频繁。具体资源的使用可分为两种情况。

1）在 Java 代码中使用资源

格式为：[＜package_name＞.]R.＜resource_type＞.＜resource_name＞

其中，package_name 为制定 R 类所在的包，如果在 Java 程序中导入 R 类所在的包则可以省略包名；resource_type 为 R 类中代表不同资源类型的子类，如 string 字符串资源；resource_name 是指资源名称，可以是 XML 文件，也可以是图片资源等，例如以下代码内容：

1. …
2. //获取指定的 TextView 组件 msg
3. TextView tv =（TextView）findViewById(R.id.msg)；
4. //设置该组件显示 string 资源中的 message 资源内容
5. Tv.setText(R.string.message)；
6. //从 drawable 资源中加载图片名为 back 的图片，并将此图片设置为当前窗口的背景
7. getWindow(0.setBackgroundDrawableResource(R.drawable.back)；
8. …

当然，有时在 Java 代码中不仅需要引用资源，有时需要访问实际的资源，这可以使用 Android 提供的 Resource 类来实现。

Resource 类提供了两个方法。

① getXxx（int id）：根据资源清单 id 来获取实际的资源。

② getAssets（）：获取所访问 assets 目录下资源的 AssetManager 对象。

如以下代码内容：

1. …
2. //首先调用 Activity 中的 getResources 方法获取资源对象
3. Resources res = getResources()；
4. //获取数组资源
5. int[] arr = res.getIntArray(R.array.students)；
6. //获取 drawable 资源
7. Drawable a = res.getDrawable(R.drawable.back)；
8. …

2）在 XML 文件中使用资源

在 XML 文件中使用资源更加简单，只要按照如下的格式访问即可：

@＜资源对应的内部类的类名＞/＜资源项名称＞

如要访问上述字符串资源中定义的"HelloWorld"字符串常量，则可以使用如下方式引用：

@string/app_name

但也有一种例外情况,当在 XML 文件中使用标识符时,这些标识符无须使用专门的资源进行定义,直接在 XML 文档中按照如下格式分配标识符即可:

@＋id/＜标识符代号＞

如上面的案例中定义的一个文本组件的语句如下:

android:id="@＋id/text"

上面的语句为这个文本组件分配了一个标识符,这个标识符在 Java 代码中可以直接使用,即使用 Activity 类中的 findViewById 这个方法就可以获取该组件;如果要在 XML 文件中获取这个组件,则可以通过资源引用的方式直接引用它,即:

@＋id/text

【任务拓展】 关于 Android 颜色值

关于颜色,读者应该了解一些理论:任何颜色理论上都可以通过红(Red)、绿(Green)及蓝(Blue)3 色合成,有时增加一个透明度 Alpha 值来表示。在 Android 颜色资源中,颜色值都用#开头,采用"透明度－红－绿－蓝"格式即"ARGB"来表示一种颜色,当对颜色的透明度没有要求时可以省略透明度,用"红－绿－蓝"格式即"RGB"来表示一种颜色。

其实,Android 颜色值可以支持 4 种格式。

1) #RGB 格式:分别指定红、绿、蓝 3 种色值来代表颜色,但只支持 0 ~ F 共 16 级颜色。

2) #ARGB 格式:分别指定透明度及红、绿、蓝 3 种色值,但不管是透明度还是红、绿、蓝 3 种色值,也只支持 0 ~ F 共 16 级。

3) #RRGGBB:分别指定红、绿、蓝 3 种色值来代表颜色,但可以支持 0 ~ FF 共 256 级颜色。

4) #AARRGGBB:分别指定透明度及红、绿、蓝 3 种色值,但不管是透明度还是红、绿、蓝 3 种色值,都支持 0 ~ FF 共 256 级。

为了方便开发者定义颜色,以下给出了颜色资源文件 colors.xml 的内容,罗列了采用 #RGB 格式给出的常用颜色值,具体如表 4-1 所列。

表 4-1 #RGB 格式常用颜色列表

颜色值	备注	颜色值	备注	颜色值	备注	颜色值	备注
FFFFFF	白色	FFFFF0	象牙色	FFFFE0	亮黄色	FFFF00	黄色
FFFAFA	雪白色	FFFAF0	花白色	FFF8DC	米绸色	FF0000	红色
C0C0C0	银色	FFE4E1	浅玫瑰色	FFE4C4	桔黄色	F0E68C	黄褐色
FFDAB9	桃色	FFD700	金色	FFC0CB	粉红色	FFB6C1	亮粉红色
FFA500	橙色	FFA07A	亮肉色	FF8C00	暗桔黄色	FF7F50	珊瑚色
FF69B4	热粉红色	FF6347	西红柿色	FF4500	红橙色	FF1493	深粉红色
FAFAD2	亮金黄色	FAF0E6	亚麻色	FAEBD7	古董白	FA8072	鲜肉色
F5DEB3	浅黄色	F4A460	沙褐色	F0FFFF	天蓝色	F5F5DC	米色
EE82EE	紫罗兰色	E9967A	暗肉色	E6E6FA	淡紫色	F0FFF0	蜜色
DEB887	实木色	DDA0DD	洋李色	DCDCDC	淡灰色	E0FFFF	亮青色

(续)

颜色值	备注	颜色值	备注	颜色值	备注	颜色值	备注
D3D3D3	亮灰色	D3D3D3	亮灰色	D2B48C	茶色	D2691E	巧可力色
A52A2A	褐色	9932CC	暗紫色	B0E0E6	粉蓝色	B8860B	暗金黄色
ADFF2F	黄绿色	ADD8E6	亮蓝色	66CDAA	中绿色	4B0082	靛青色
9400D3	暗紫罗兰色	9370DB	中紫色	90EE90	亮绿色	8B0000	暗红色
8B008B	暗洋红	87CEEB	天蓝色	808080	灰色	8A2BE2	紫罗兰蓝色
808000	橄榄色	800080	紫色	800000	粟色	7FFFD4	碧绿色
7FFF00	黄绿色	7CFC00	草绿色	696969	暗灰色	778899	亮蓝灰
008000	绿色	708090	灰石色	6B8E23	深绿褐色	0000FF	蓝色
696969	暗灰色	40E0D0	青绿色	00FFFF	浅绿色	00FFFF	青色
00008B	暗蓝色	000080	海军色	000000	黑色	FFFFFF	白色

任务2 数组资源的使用

【任务描述】

同 Java 一样,Android 中也允许使用数组。但是在 Android 中,不推荐在 Java 程序中定义数组,而是推荐使用数组资源文件来定义数组。下面对数组资源进行详细介绍。

数组资源文件位于 res \ values 目录下,资源文件一般以 arrays.xml 为文件名来定义数组。定义数组时,根元素也是 <resources ></resources > 标记,在该元素中,包括以下3个子元素。

1) < array > 子元素:用于定义普通类型的数组。
2) < integer – array > 子元素:用于定义整数数组。
3) < string – array > 子元素:用于定义字符串数组。

本任务是创建数组资源文件并在界面中引用。

【任务实施】

1) 创建 chap04_2 工程项目,在/res/values/目录中创建 arrays.xml 资源文件,用数组资源定义数组的方法对资源文件进行编辑,其程序内容如下:

```
1.   <?xml version = "1.0" encoding = "utf – 8"?>
2.   <resources >
3.       <!--定义一个 Drawable 数组-->
4.       <array name = "plain_arr" >
5.           <item > @color/c1 </item >
6.           <item > @color/c2 </item >
7.           <item > @color/c3 </item >
8.           <item > @color/c4 </item >
9.           <item > @color/c5 </item >
10.          <item > @color/c6 </item >
```

```
11.        <item>@color/c7</item>
12.        <item>@color/c8</item>
13.        <item>@color/c9</item>
14.     </array>
15.     <!--定义字符串数组-->
16.     <string-array name="string_arr">
17.        <item>@string/c1</item>
18.        <item>@string/c2</item>
19.        <item>@string/c3</item>
20.        <item>@string/c4</item>
21.        <item>@string/c5</item>
22.        <item>@string/c6</item>
23.        <item>@string/c7</item>
24.        <item>@string/c8</item>
25.        <item>@string/c9</item>
26.     </string-array>
27.     <!--定义字符串数组-->
28.     <string-array name="books">
29.        <item>Java 面向对象编程</item>
30.        <item>JSP 编程</item>
31.        <item>Android 基础开发教程</item>
32.     </string-array>
33. </resources>
```

2）定义了数组资源之后，既可以在 XML 文件中使用这些数组资源，也可以在 Java 程序中使用这些数组资源。下面在界面布局文件 main.xml 中定义一个 ListView 数组，将 android:entries 属性指定为一个数组，其界面布局文件对应的程序如下所示。

```
1.  <?xml version="1.0" encoding="utf-8"?>
2.  <LinearLayout xmlns:android="http://schemas.android.com/apk/res/android"
3.      android:orientation="vertical"
4.      android:layout_width="fill_parent"
5.      android:layout_height="fill_parent"
6.      android:gravity="center_horizontal">
7.      <!--使用字符串资源、尺度资源-->
8.  <TextView
9.      android:layout_width="wrap_content"
10.     android:layout_height="wrap_content"
11.     android:text="@string/app_name"
12.     android:gravity="center"
13.     android:textSize="@dimen/title_font_size"/>
14.     <!--定义一个 GridView 组件,使用尺度资源中定义的长度来指定水平间距、垂直间距-->
15. <GridView
```

```
16.        android:id = "@+id/grid01"
17.        android:layout_width = "wrap_content"
18.        android:layout_height = "wrap_content"
19.        android:horizontalSpacing = "@dimen/spacing"
20.        android:verticalSpacing = "@dimen/spacing"
21.        android:numColumns = "3"
22.        android:gravity = "center" >
23.    </GridView>
24.    <!-- 定义 ListView 组件,使用了数组资源 -->
25.    <ListView
26.        android:layout_width = "wrap_content"
27.        android:layout_height = "wrap_content"
28.        android:entries = "@array/books"   />
29. </LinearLayout>
```

3)接下来,在程序中直接使用资源文件中定义的数组,其相应代码如下。

```
1.  public class ArrayResActivity extends Activity
2.  {
3.      //获取系统定义的数组资源
4.      String[] texts;
5.      public void onCreate(Bundle savedInstanceState)
6.      {
7.          super.onCreate(savedInstanceState);
8.          setContentView(R.layout.main);
9.          texts = getResources().getStringArray(R.array.string_arr);
10.         //创建一个 BaseAdapter 对象
11.         BaseAdapter ba = new BaseAdapter()
12.         {
13.             public int getCount()
14.             {
15.                 //指定一共包含9个选项
16.                 return texts.length;
17.             }
18.             public Object getItem(int position)
19.             {
20.                 // 返回指定位置的文本
21.                 return texts[position];
22.             }
23.             public long getItemId(int position)
24.             {
25.                 return position;
26.             }
```

```
27.            //重写该方法,该方法返回的 View 将作为 GridView 的每个格子
28.            @Override
29.            public View getView(int position,
30.                    View convertView, ViewGroup parent)
31.            {
32.                TextView text = new TextView(ArrayResActivity.this);
33.                Resources res = ArrayResActivity.this.getResources();
34.                //使用尺度资源来设置文本框的高度、宽度
35.                text.setWidth((int) res.getDimension(R.dimen.cell_width));
36.                text.setHeight((int) res.getDimension(R.dimen.cell_height));
37.                //使用字符串资源设置文本框的内容
38.                text.setText(texts[position]);
39.                TypedArray icons = res.obtainTypedArray(R.array.plain_arr);
40.                //使用颜色资源来设置文本框的背景色
41.                text.setBackgroundDrawable(icons.getDrawable(position));
42.                text.setTextSize(20);
43.                return text;
44.            }
45.        };
46.        GridView grid = (GridView) findViewById(R.id.grid01);
47.        grid.setAdapter(ba);  //为 GridView 设置 Adapter
48.    }
49. }
```

上面程序中的粗体字代码是使用数组资源的关键代码。运行上面的程序后可看到如图 4-9 所示的界面。

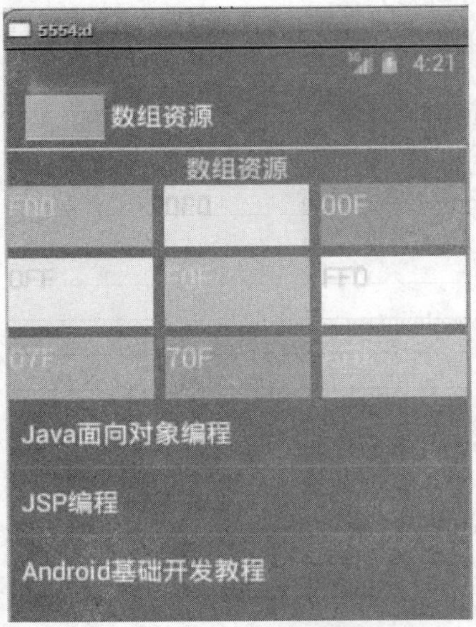

图 4-9　应用运行界面

任务3　Drawable 资源的使用

【任务描述】

Drawable 资源是保存在 res/drawable 目录下的，它是 Android 应用中使用最频繁的资源，不仅可以直接使用 .png、.jpg、.gif、.jpeg 等图片作为资源，也可以使用多种 XML 文件作为资源。只要将 XML 文件编译成 Drawable 子类的对象，即可作为 Drawable 资源使用。

开发者可以使用 Drawable 资源类型确定一个简单的带颜色矩形，并可以将之运用到屏幕中。具体做法是在 drawable 的目录下定义一个 XML 资源文件，这个资源文件可用以绘制 Drawable 资源，绘制时需使用其相应的 <drawable> 标识，并使用"键/值"对的方式来定义。

本任务将基于 Drawable 资源为读者实现一个功能强大的按钮。

【任务实施】

1）创建 chap04_3 工程项目，在生成的项目的 res/drawable_mdpi 目录下将 red.png 等 3 个图片添加进来，并在此目录下创建一个 XML 文件 button_selector.xml，其相应代码如下所示。

```
1.  <?xml version = "1.0" encoding = "utf-8"?>
2.  <selector xmlns:android = "http://schemas.android.com/apk/res/android">
3.      <item android:state_pressed = "true" android:drawable = "@drawable/red"></item>
4.      <item android:state_pressed = "false" android:drawable = "@drawable/purple"></item>
5.  </selector>
```

上述资源文件使用了 <selector/> 元素定义了一个 stateListDrawable 对象，它可以用来定义按钮组件在不同的状态下使用不同的颜色或 Drawable 资源。如上述代码定义了当按钮被按下时显示 red 背景图片，当松开鼠标时，背景又变成了 purple 背景。

2）打开 main.xml 布局文件，在其中添加了两个按钮，第一个按钮直接使用 blue 背景图片，而第二个按钮使用了 Drawable 对象"button_selector/xml"，其相应代码如下。

```
1.  <LinearLayout xmlns:android = "http://schemas.android.com/apk/res/android"
2.      xmlns:tools = "http://schemas.android.com/tools"
3.      android:layout_width = "match_parent"
4.      android:layout_height = "match_parent">
5.      <Button
6.          android:layout_width = "wrap_content"
7.          android:layout_height = "wrap_content"
8.          android:background = "@drawable/blue"
9.          android:text = "普通按钮"
10.         android:textSize = "10pt"
11.         android:textColor = "#ffffff"/>
12.     <LinearLayout
13.         android:layout_width = "fill_parent"
```

14.	android:layout_height = "fill_parent"
15.	android:orientation = "horizontal" >
16.	<Button
17.	android:layout_width = "wrap_content"
18.	android:layout_height = "wrap_content"
19.	android:background = "@drawable/button_selector"
20.	android:text = "背景可以变化的按钮"
21.	android:textSize = "10pt"
22.	android:textColor = "#ffffff" />
23.	</LinearLayout>
24.	</LinearLayout>

运行上述应用程序后，可见如图4-10所示的运行界面。

上面两个按钮中，第一个按钮的背景颜色是固定的，第二个按钮的背景图片是可以切换的，即当把鼠标指针放在上述按钮的第二个按钮即"背景可以变化的按钮"上并按住鼠标左键时，按钮的背景图片被切换成了红色。

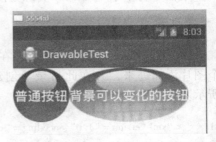

图4-10 应用运行界面

【知识学习】图片资源

Android支持多种图片格式。

（1）便携式网络图像格式

这是一种无损格式，扩展名为.png，也是Android推荐使用的格式。

（2）九格拉伸图像

手机屏幕拥有各种不同的尺寸。如果图像可以根据手机不同的屏幕尺寸、方向或者文字长度的不同而自动调整大小，它将非常方便适用。这可以极大节约开发人员为各种不同屏幕尺寸创建图片的时间。而九格拉伸图像就可以达到这一目的。它可以使用draw9patch工具从.png文件创建，它位于Android SDK的tools目录下。

（3）联合图像专家组格式及图形交换格式

前者格式为JPG或JPEG，是Android可以接受的格式，但是有损的；后者为GIF格式，这种格式在Android中不推荐使用。

将图片按照Android资源的命名规则进行命名后，只要将图片放入res/drawable-xx目录下，Eclipse就会在R资源清单类中自动生成该资源的索引。此时，就可以在Java类或XML中使用前面项目中介绍的格式进行资源的访问了。

【问题研讨】在Java代码中使用图像资源

图像资源是Android中最常见的资源形式，因此前面介绍的资源的使用方式，包括在Java代码和XML文件中的使用方式也适合图像资源。这里重点介绍在程序中使用图像资源的注意点。

图像资源是另一种Drawable，被称为BitmapDrawable。很多情况下，开发者仅需要图像的资源 id 就可以设置用户界面空间的属性。例如，如果把图片文件 jsp.png 放入/res/drawable/目录下，并向布局中加入 ImageView 控件，其名字为"photo"，其具体程序如下：

```
1.  import android.widget.ImageView;
2.  …
3.  ImageView iv = (ImageView)findViewById(R.id.photo);//获取界面中的 ImageView 控件
4.  //将图像 jsp 赋予 ImageView 控件的 imageResource 属性即可显示图像
5.  iv.setImageResource(R.drawable.jsp);
6.  …
```

如果需要直接访问 BitmapDrawable 对象，可以用如下代码实现：

```
1.  import android.graphics.drawable.BitmapDrawable;
2.  …
3.  BitmapDrawable bmpd = (BitmapDrawable)getResources().getDrawable(R.drawable.jsp);
4.  int iBitmapHeightInPixela = bmpd.getIntrinsicHeight();
5.  int iBitmapWeidthInPixela = bmpd.getIntrinsicWidth();
6.  …
```

如果使用的图像资源是九格拉伸图像（如 serv.9.png），则对 getDrawable() 的调用将返回一个 NinePatchDrawable 对象，而不是一个 BitmapDrawable 对象。

```
1.  import android.graphics.drawable.NinePatchDrawable;
2.  …
3.  NinePatchDrawable npd =
4.     (NinePatchDrawable)getResources().getDrawable(R.drawable.serv);
5.  int iBitmapHeightInPixela = npd.getIntrinsicHeight();
6.  int iBitmapWeidthInPixela = npd.getIntrinsicWidth();
7.  …
```

请读者赶快试试吧。

【任务拓展】

读者对布局资源 Layout 应该不陌生，因为前面的内容已经介绍了 Android 布局资源的知识了。因此这里不会做详细讲解，仅仅对 Layout 布局资源做个简单的总结。

Layout 布局资源位于 Android 的/res/layout/目录下，布局文件的根元素是前述的各种类型的布局管理器，如线性布局管理器、表格布局管理器、相对布局管理器等。

对布局资源的使用也分为两种情况，即在 XML 文件中引用和在 Java 代码中引用或直接使用。在 XML 文件下引用的格式是：

@[<package_name:>]layout/文件名

在 Java 代码中引用的格式是：

@[<package_name:>]R.layout.<文件名>

模块二 Android 系统资源及 assets 资源的使用

【模块描述】

模块一中介绍的资源为 Android 的 res 目录资源，除此之外还有 Android 系统资源和 assets 资源。本模块主要为读者讲解 Android 系统资源及 assets 资源的访问方式。

知识点	技能点
➢ Android 资源分类 ➢ res 目录资源与系统资源及 assets 资源的区别	➢ Android 系统资源的访问及使用方法 ➢ assets 资源的使用方式

任务 1 Android 系统资源的访问和使用

【任务描述】

在 Android 应用中，除了可以使用自定义的资源外，应用程序还可以访问系统提供的各种丰富的资源（Android SDK 提供的系统资源）。当成功安装 Android SDK 后，这个资源存放在/platform/ < platform_version >/data/res 目录下。其中， < platform_version > 为 Android SDK 对应的平台，本书采用的是 4.2 版本，对应的是 android - 17 平台，其部分资源目录如图 4-11 所示。

图 4-11 Android 系统资源（部分）

Android 系统资源的使用格式是：android. R. 资源类型. 资源 id。注意"android"是小写的。在 Eclipse 的编辑器中，可以通过编码提示引用系统资源。

【任务实施】

1）创建 chap04_4 工程项目，在布局文件中拖入一个 ImageView 控件 iv，其具体 XML 代码如下。

```
1.    < ImageView
2.        android:id = " @ + id/iv"
```

```
3.          android:layout_width = "wrap_content"
4.          android:layout_height = "wrap_content"
5.          android:layout_alignParentLeft = "true"
6.          android:layout_alignParentTop = "true"
7.          android:layout_marginLeft = "138dp"
8.          android:layout_marginTop = "178dp"
9.          />
```

2）打开 Activity，首先获取 ImageView 控件，然后引用系统资源，即将 Drawable 资源放入 ImageView 控件中显示，其修改后的代码如下。

```
1.  …
2.  protected void onCreate( Bundle savedInstanceState) {
3.      super. onCreate( savedInstanceState);
4.      setContentView( R. layout. main);
5.      ImageView iv = ( ImageView) this. findViewById( R. id. iv);
6.      iv. setImageResource( android. R. drawable. stat_sys_phone_call_forward);
7.  }
8.  …
```

运行 chap04_4 应用，可以看到成功引用了系统资源并将其显示在界面中，如图 4-12 所示。

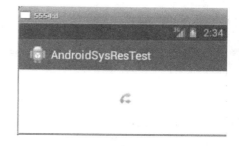

图 4-12 系统资源引用

任务 2 assets 资源的使用

【任务描述】

前面已经介绍过，Android 应用下的 assets 目录也可以存放资源，但放在这个目录下的资源无法在 R. java 中生成资源 id。所以，如果要使用 assers 目录下的资源，需要直接使用资源名称访问。

【任务实施】

1）创建 chap04_5 工程项目，在布局文件中拖入一个 TextView 控件 tv，具体的 XML 代码略。

2）在 assets 目录下创建一个 "content. txt" 文件，在里面随便输入一些英文字符，比如这里将 Activity 中的代码复制到这个文件中，创建完后如图 4-13 所示。

3）打开 Activity，修改其中的 onCreate 方法，改后的代码如下。

```
1.  protected void onCreate( Bundle savedInstanceState) {
2.      super. onCreate( savedInstanceState);
3.      setContentView( R. layout. main);
4.      TextView tv = ( TextView) this. findViewById( R. id. show);
5.      try {
```

```
6.              InputStream is = this.getAssets().open("content.txt");
7.              byte[] buffer = new byte[1024];
8.              int c = is.read(buffer);
9.              String s = new String(buffer,0,c);
10.             tv.setText(s);
11.         }catch(IOException e){
12.             //TODO Auto-generated catch block
13.             e.printStackTrace();
14.         }
15.     }
```

运行 chap04_5 应用，可看到界面中显示的内容是从 assets 目录下的 content.txt 文件中读取的，如图 4-14 所示。

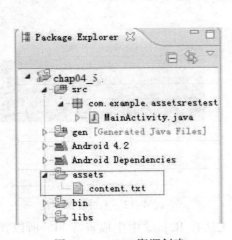

图 4-13 assets 资源创建

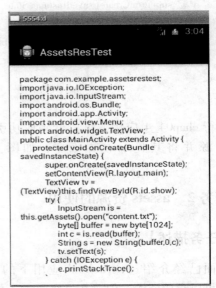

图 4-14 访问 assets 目录资源

【练习】

对项目三中的习题 2. "模仿实现 QQ 登录布局界面"进行修改，将布局文件中用到的颜色、尺寸、字符串、图片等相关资源都放入资源文件中，要求如下：

1）将界面中所有显示的文字放入 strings.xml 资源文件中，并在布局文件中对其正确引用；

2）在 res 文件夹中创建一个 colors.xml 资源文件，将界面中所有的"颜色"放入 colors.xml 资源中，并在布局文件中对其正确引用；

3）在 res 文件夹中创建一个 dimens.xl 资源文件，将"界面尺寸"放入 dimens.xml 资源文件中，并在布局文件中对其正确引用；

4）讨论上述这种做法相比项目三中的做法，各有什么优点或好处。

项目五 使用 Intent 实现界面跳转

Android 应用程序主要由 4 个部分组成,分别为活动(Activity 组件)、服务(Service 组件)、内容提供者(Content Provider)和广播接收者(BroadcastReceiver 组件)。其中,Activity 组件是 Android 应用的入口,是 Android 初级程序员必须掌握的组件之一。本项目重点讲解 Activity 组件的概念、建立、配置及基于 Intent 实现界面跳转的方法,在此基础上还给出了使用 Bundle 在不同 Activity 之间进行数据交换的方法,并讨论了 Activity 的生命周期。

【知识目标】

- Activity 的功能和作用
- Activity 的建立、启动、配置、关闭等操作
- Activity 的回调机制
- Activity 的生命周期
- Activity 的 4 种加载模式
- Activity 之间数据交换的方法
- Intent 的含义及用法
- Bundle 的用法
- Android 事件监听处理机制
- Android 常用事件及事件监听器
- Handler 消息传递机制

【模块分解】

- 创建、启动及配置 Activity
- 会使用 Bundle 在不同的 Activity 之间进行数据交换
- Android 事件监听处理

模块一 Activity 组件的创建/启动/配置

【模块描述】

Activity 组件的创建必须直接或间接继承 android.app.Activity 类,或者继承其子类,根据业务需求可以选择覆盖其中的合适的方法。在使用 Eclipse 作为 Android 应用的开发环境创建一个 Android 应用时,将会创建一个默认的 Activity。下面将通过继承 ListActivity 类实现游戏列表界面。

知识点	技能点
➤ Activity 的概念及生命周期 ➤ Activity 的创建、启动、关闭、配置方法 ➤ Activity 类及其子类的特征 ➤ ListActivity 子类的使用场合 ➤ Intent 的含义及使用	➤ 使用 Eclipse ADT 可视化创建 Activlty 的方法 ➤ 使用 Eclipse ADT 可视化配置 Activity 的方法 ➤ 继承 ListActivity 子类实现 Activity ➤ 使用 Intent 实现 Activity 之间的跳转

任务1　Activity 组件的创建及配置

【任务描述】

在 Eclipse ADT 开发环境下通过继承 ListActivity 子类来定义 Activity，可视化创建、启动、配置 Activity，实现游戏列表界面。

【任务实施】

1）创建一个 Android 应用，项目名称为"chap05_1"，应用的名字为"GameManager"。连续单击 Next 按钮到如图 5-1 所示的界面。

图 5-1　选择创建的 Activity 类型

2）如果在上图中勾选了"Create Activity"复选框，则会进入创建 Activity 界面，如图 5-2 所示。

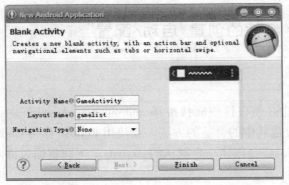

图 5-2　创建 Activity 界面

3）在上图中设置 Activity 及布局文件参数，单击 Finish 按钮，则创建项目结束。

4）在 res/layout 目录下打开 gameslist.xml 文件，添加一个 TextView 组件，具体代码如下所示。

```xml
1.  <?xml version = "1.0" encoding = "utf-8"?>
2.  <TextView
3.      android:layout_width = "match_parent"
4.      android:layout_height = "40sp"
5.      android:id = "@ + id/gamesItem"
6.      android:paddingLeft = "6dip"
7.      android:textSize = "22sp"
8.      android:textStyle = "normal"
9.      android:background = "#ff0000"
10.     xmlns:android = "http://schemas.android.com/apk/res/android" >
11. </TextView>
```

第 3 行指定这个列表中的 TextView 的宽度为父容器的宽度，第 4 行代码指定列表中每行的高度为 40sp，第 6 行指定列表的左间距为 6dip，第 7 行指定列表中的字体大小为 22sp，第 9 行指定列表的背景颜色为红色。

5）打开 GamesActivity.java，修改这个类的父类，使之继承 ListActivity 类，其具体代码如下所示。

```java
1.  public class GamesActivity extends ListActivity {
2.      private String[] games = {"俄罗斯方块","超级玛丽","变形小狗","忍者","卡哇伊狗狗"};
3.      protected void onCreate(Bundle savedInstanceState) {
4.          super.onCreate(savedInstanceState);
5.          this.setListAdapter(new ArrayAdapter<String>(this, R.layout.gamesitem, games));
6.      }
7.      public boolean onCreateOptionsMenu(Menu menu) {
8.          //Inflate the menu; this adds items to the action bar if it is present.
9.          getMenuInflater().inflate(R.menu.games, menu);
10.         return true;
11.     }
12. }
```

上面的 Activity 通过继承 ListActivity 子类可以实现简单的列表界面。第 2 行代码用于定义了一个一维数组 games，第 5 行则定义了一个 ArrayAdapter 数组适配器，这个适配器需要 3 个参数，上下文"this"（当前类 ListActivity 之对象）、XML 布局文件在 R 文件中对应的 int 类型常量和 int 型数组，然后调用 ListActivity 子类中的 setListAdapter 方法为这个列表设置 Adapter。当 Andriod 应用运行时，就可以显示列表的界面了，如图 5-3 所示。

图 5-3　chap05_1 运行界面

Android 应用要求应用程序的 Activity 组件必须进行配置，只有这样程序才可以运行。但读者发现，前面所有的案例都没有进行配置，为什么可以运行呢？

这是因为使用 Eclipse 会在创建 Acitvity 时自动配置。打开上面案例的一个 XML 文件 AndroidManifest.xml。这个文件位于应用的根目录下。双击这个文件，可见到如图 5-4 所示的图。

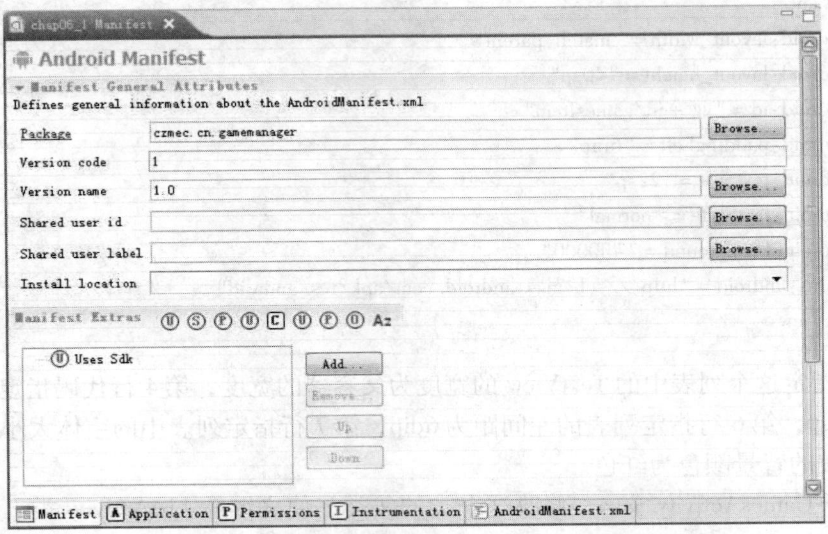

图 5-4 AndroidManifest.xml 文件

上图有 5 个选项卡，分别为"Mainfest""Application""Permissions""AndroidManifest.xml"及"Instrumentation"，下面对常用的前 4 个选项卡的功能进行介绍。

① Mainfest 选项卡。用以可视化设置所属的包、Manifest 版本号及版本名称、应用所使用的 Android SDK 的最低版本等。

② Application 选项卡。用以可视化设置应用的主题名称、应用的图标等。

③ Permissions 选项卡。通过之前的内容读者已经知道，Android 应用的相互调用需要对应的权限才行。Permissions 选项卡用以可视化设置应用程序的许可证、权限等。

④ AndroidManifest.xml 选项卡。用以直接编辑相关应用配置，包括 Activity 的配置，如图 5-5 所示。

上图展示了 Activity 的配置，即在 <application.../> 节点中添加一个子节点 <activity.../>，同时指定 activity 节点的相关属性。下面对 activity 节点的常用属性进行介绍。

① name 属性：指定该 Activity 的实现类的名称。

② label 属性：指定 Activity 运行时显示的名字；图 5-5 中设置为"我的游戏列表"，当这个 Activity 运行时将会被显示出来。当然 android.app.Activity 类也定义了大量的方法，可以通过编程的方式修改或获取 Activity 的配置信息。如 lable 这个属性的配置，可以通过使用 android.app.Activity 类中的 setTitle 方法设置，即修改 GamesManager 类，在其中加入如下代码即可：this.setTitle("我的游戏列表")。

③ exported 属性：用来设置该 Activity 是否允许被其他应用调用，如果是"true"，则允许被其他应用程序调用。

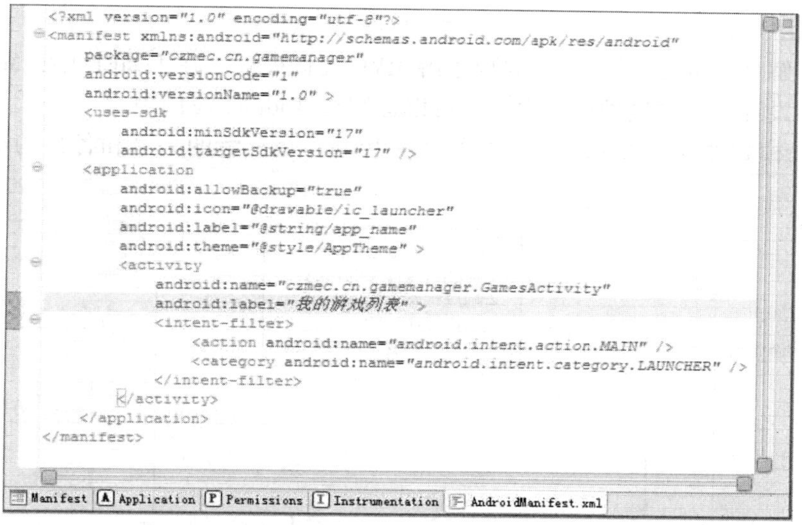

```xml
<?xml version="1.0" encoding="utf-8"?>
<manifest xmlns:android="http://schemas.android.com/apk/res/android"
    package="czmec.cn.gamemanager"
    android:versionCode="1"
    android:versionName="1.0" >
    <uses-sdk
        android:minSdkVersion="17"
        android:targetSdkVersion="17" />
    <application
        android:allowBackup="true"
        android:icon="@drawable/ic_launcher"
        android:label="@string/app_name"
        android:theme="@style/AppTheme" >
        <activity
            android:name="czmec.cn.gamemanager.GamesActivity"
            android:label="我的游戏列表" >
            <intent-filter>
                <action android:name="android.intent.action.MAIN" />
                <category android:name="android.intent.category.LAUNCHER" />
            </intent-filter>
        </activity>
    </application>
</manifest>
```

图 5-5 AndroidManifest.xml 选项卡

④ launchMode 属性：用来指定该 Activity 的加载模式。该属性有 standard、singleTop、singleTask 及 singleInstance 4 种加载模式，在后面将会详细介绍。

在 < activity > 节点里，有个子节点 < intent - filter…/ >，其相应代码如下所示：

```
1.    < activity
2.        android:name = "czmec. cn. gamemanager. GamesActivity"
3.        android:label = "我的游戏列表" >
4.        < intent - filter >
5.            < action android:name = "android. intent. action. MAIN" />
6.            < category android:name = "android. intent. category. LAUNCHER" />
7.        </ intent - filter >
8.    </ activity >
```

上述代码配置了一个 activity，并在这个 activity 中配置了一个 < intent - filter > 节点，这个节点的作用是用于指定该 Activity 可以响应的 Intent。此处即指定了该 Activity 作为应用程序的入口。

< intent - filter > 节点中比较常用的两个子节点是 action 和 category。这两个节点的唯一属性都是 android:name。其中，action 节点的 android:name 用来指定该 Activity 所接收的动作，上述配置中的 android. intent. action. MAIN 表示程序的入口，即应用被运行后第一个被调用的 Activity，这就是当运行 chap05_1 时将自动调用 GamesManager 类的原因。< category > 子节点表示 Activity 的种类，其中，android. intent. category. LAUNCHER 表示该 Activity 被显示在 Android 系统的最顶层。

【知识学习】深入理解 Activity 组件

（1）Activity 类及其概念

Activity 是 Android 系统中的四大组件之一，是 Android 应用的重要组成单元之一，可以用于显示 View。

在深入了解 Activity 之前，读者先要了解 MVC 设计模式。在 JavaEE 中，MVC 设计模式是一种经典的设计模式，这种设计模式也可以运用在 Android 应用中。

MVC 设计模式即 M—模型、V—视图、C—控制器，三者的关系和各自的功能如图 5-6 所示。

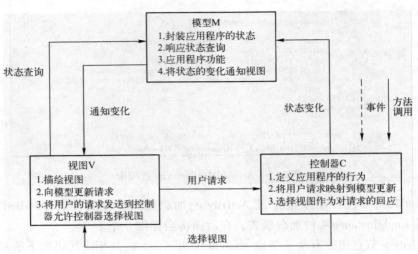

图 5-6　MVC 设计模式

MVC 设计模式共分为 3 层。

1）模型（Model）：代表应用程序的业务逻辑状态。Model 是应用程序的主体部分，所有的业务逻辑都应该写在这里。在 Android 中，Model 层与 JavaEE 中的变化不大，如对数据库的操作、对网络等的操作都放在该层（但不是说它们都放在同一个包中，可以分开放，但它们统称为 Model 层）。

2）视图（View）：提供可交互的用户界面，向客户展示模型中的数据，是应用程序中负责生成用户界面的部分，也是在整个 MVC 架构中用户唯一可以看到的一层，接收用户输入，显示处理结果；在 Android 应用中一般采用 XML 文件里的界面的描述，使用的时候可以非常方便地引入，当然也可以使用 JavaScript + HTML 等方式作为 View。

3）控制器（Control）：响应客户请求，根据客户的请求来操作模型并把模型的响应结果经由视图呈现给客户端。Android 的控制层的重任就要落在应用广泛的 Activity 的肩上了，所以建议开发者不要在 Activity 中写太多的代码，尽量将功能实现交给 Model 业务逻辑层处理。

好了，在介绍过 Android 应用开发中的 MVC 架构后，读者就可以明确地知道，在 Android 中 Activity 主要是用来做控制的，它可以选择要显示的 View，也可以从 View 中获取数据，然后把数据传给 Model 层进行处理，最后再来显示出处理结果。有过 Web 开发经验的读者对 Servlet 概念应该比较熟悉，实际上，Activity 对于 Android 应用的作用有点类似于 Servlet 对于 Web 应用的作用——一个 Web 应用通常由多个 Servlet 组成，那么一个 Android 应用通常也由多个 Activity 组成。

那么到底 Activity 是什么呢？

从表面上看,Activity 是 Android 应用程序的一个图形用户界面,用户通过这个界面可以操作 Android 应用;对于开发者而言,Activity 是程序的一个入口,是一个具有一定编程规范的 Java 类,必须直接或间接继承 android.app.Activity 类。

在前面模块的案例中,读者看到一个 Android 应用仅仅包含一个 Activity,但这在实际应用中是不可能的,实际应用中往往包含多个 Activity,不同的 Activity 向用户呈现不同的操作界面。

1)一个 Activity 是一个应用程序组件,提供一个屏幕,用户可以用来交互以完成某项任务,例如拨号、拍照、发送 E-mail、看地图。每一个 Activity 被给予一个窗口,在其中可以绘制用户接口。窗口通常充满屏幕,但也可以小于屏幕而浮于其他窗口之上。

2)一个应用程序通常由多个 Activity 组成,它们通常是松耦合关系。通常,一个应用程序中被指定为"main"的 Activity 是第一次启动应用程序的时候呈现给用户的那个 Activity。一个 Activity 可以启动另一个 Activity 以完成不同的动作。每一次一个 Activity 启动,前一个 Activity 就停止了,但是系统保留 Activity 在一个栈上(Back Stack)。当一个新 Activity 启动,它被推送到栈顶,取得用户焦点。Back Stack 符合简单"后进先出"原则,所以,当用户完成当前 Activity 后单击 back 按钮,它被弹出栈(并且被摧毁),然后之前的 Activity 恢复。

(2)Activity 子类及其使用注意事项

开发者创建一个 Activity 组件,必须直接或间接继承 android.app.Activity 基类。当然,在不同的应用场合下,有时也要求继承不同的 Activity 子类。比如,如果程序开发者需要在界面中定义一个列表来显示数据,那么可以在定义 Activity 组件时继承 ListActivity;如果应用程序界面需要实现标签页的效果,则可以在定义 Activity 时继承 TabActivity 来实现。

Activity 及其对应的子类如图 5-7 所示。

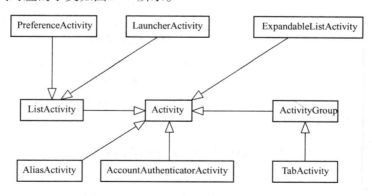

图 5-7 Activity 组件及其对应的子类图

从上图中可以看到,有几个非常重要的 Activity 子类需要特别注意,如 ListActivity、PreferenceActivity 和 TabActivity 等。ListActivity 可以用来实现列表界面,PreferenceActivity 可以实现参数设置及存储界面,TabActivity 可以实现 Tab 界面,ExpandableListActivity 可以实现可展开列表界面。

开发者要创建一个 Activity 组件时必须直接或间接继承 android.app.Activity 类或子类,这就必须覆盖父类中的某些方法。Activity 类中有几个重要的方法。

1) onCreate 方法

onCreate 方法用来初始化 Activity，常常被子类覆盖，用来初始化当前的 Activity。如前面的案例中创建 Activity 类，都继承了 android. app. Activity 类且覆盖了 onCreate 方法，如下面的代码所示。

```
1.    public class MainActivity extends Activity {
2.        public void onCreate(Bundle savedInstanceState) {      //重写 onCreate 方法
3.            super.onCreate(savedInstanceState);
4.            setContentView(R.layout.XXX);
5.            …
6.        }
7.        …
8.    }
```

2) onPause 方法

onPause 方法在用户离开当前 Activity 时使用，通常也被子类覆盖，用来实现有任何变化的处理逻辑。

3) setContentView 方法

setContentView 方法初学者经常会用到，其作用是在当前 Activity 中指定布局资源文件，这个布局文件通常是使用 res/layout 目录下的某个 XML 资源文件，这个资源文件称为布局文件。如上面代码中的第 4 行，就是通过调用其父类 Activity 中 setContentView 方法，指定该 Activity 显示的布局是 res/layout 目录下定义的 XXX. xml 资源文件。

4) findViewById 方法

大多数的 Android 应用中，UI 组件都是在 XML 布局资源文件中定义的，如 main. xml。如果在 Activity 中需要对布局文件中定义的某个 UI 组件进行编程，则可以通过 findViewById 方法获得与 id 值对应的 UI 组件对象，这点在前面介绍的很多例子中都有应用。这里需要强调的是，当使用这个方法获取到 UI 后一定要进行强制类型转换，将获得的抽象的 UI 对象转换成对应的组件类型，如 TextView tv = (TextView)findViewById(R.id.show)。

除了这 4 个方法外，Activity 中还定义了大量的方法，读者可以查阅 Android 4.2 API 帮助文档，这里不再详细介绍。

任务2　使用 Intent 启动 Activity 实现界面跳转

【任务描述】

一个 Android 应用通常包括多个界面，也就包括多个 Activity，但只有一个 Activity 会作为应用的主入口，其他的 Activity 通常是由主入口 Activity 启动的，或者由主入口 Activity 启动的 Activity 来启动。如何设置一个 Activity 为主入口，任务 1 中已经进行了详细的介绍。本任务主要介绍使用 Intent 方式启动其他 Activity 的方法。

【任务实施】

1) 创建一个 Android 应用，项目名称为"chap05_2"，应用的名字为"IntentTest"。

连续单击 Next 按钮，设置 Activity 名字为 FirstActivity，设置布局文件名字为 main，应用创建完成。

2）编辑主窗口的布局文件 main.xml，其相应代码如下。

```
1.  <LinearLayout xmlns:android = "http://schemas.android.com/apk/res/android"
2.      android:layout_width = "fill_parent"
3.      android:layout_height = "fill_parent"
4.      android:orientation = "vertical" >
5.      <TextView
6.          android:layout_width = "fill_parent"
7.          android:layout_height = "wrap_content"
8.          android:text = "你好,这是第一个界面 Activity" />
9.      <Button
10.         android:id = "@ + id/bn"
11.         android:layout_width = "wrap_content"
12.         android:layout_height = "wrap_content"
13.         android:text = "Button" />
14. </LinearLayout>
```

3）创建第 2 个窗口的 Activity。选中应用的 src 目录下的包，单击鼠标右键，在弹出的快捷菜单中选择 New→Other 命令，如图 5-8 所示。

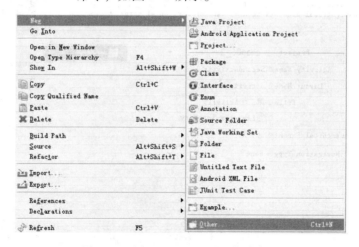

图 5-8　选择 Activity 所需的菜单命令

4）此时弹出的界面如图 5-9 所示。

5）选中上图中的 Android Activity 选项，单击 Next 按钮，可打开如图 5-10 所示的界面。

6）选中上图中的 Blank Activity 选项，单击 Next 按钮，可弹出如图 5-11 所示的界面。

7）对上图中参数进行设置后，单击 Next 按钮，可弹出如图 5-12 所示的界面。

勾选上图中的复选框，如勾选 AndroidManifest.xml 复选框，是指在生成 SecondActivity 后，自动在 AndroidManifest.xml 文件中进行 SecondActivity 配置，即在 <application> 节点中添加一个 <activity> 节点。如果不勾选，则需要开发者手动打开这个文件并添加相关代码。

如要自动生成 SecondActivity 需配置的代码如下所示：

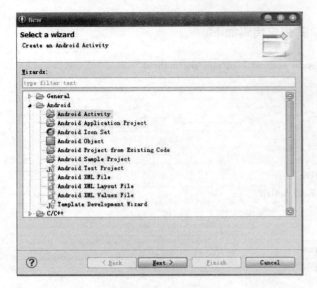

图 5-9　创建 Activity 界面（一）

图 5-10　创建 Activity 界面（二）

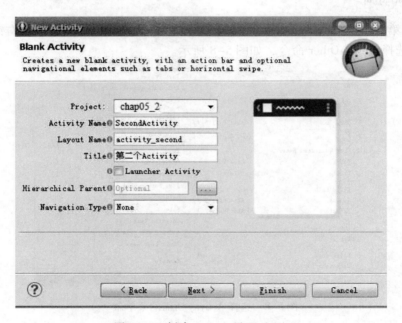

图 5-11　创建 Activity 界面（三）

```
1.    <activity
2.          android:name = "czmec.cn.intenttest.SecondActivity"
3.          android:label = "@string/title_activity_second" >
4.    </activity>
```

8) 打开生成的 second.xml 布局文件，增加两个 Button 组件，其相应的代码如下：

图 5-12 创建 Activity 界面（四）

```
1.  <?xml version = "1.0" encoding = "utf-8"?>
2.  <LinearLayout xmlns:android = "http://schemas.android.com/apk/res/android"
3.      android:layout_width = "fill_parent"
4.      android:layout_height = "fill_parent"
5.      android:orientation = "vertical" >
6.      <Button
7.          android:id = "@+id/previous"
8.          android:layout_width = "wrap_content"
9.          android:layout_height = "wrap_content"
10.         android:text = "Previous" />
11.     <Button
12.         android:id = "@+id/Close"
13.         android:layout_width = "wrap_content"
14.         android:layout_height = "wrap_content"
15.         android:text = "Close" />
16. </LinearLayout>
```

9）打开 FirstActivity 源码，在 onCreate 方法中首先获取 bn 按钮组件，为这个组件添加事件监听器，并在 onClick 单击事件中添加一个 Intent，用来启动 SecondActivity，其相应的代码如下：

```
1.  …
2.  protected void onCreate(Bundle savedInstanceState) {
```

```
3.          super.onCreate(savedInstanceState);
4.          setContentView(R.layout.main);
5.          this.setTitle("这是主窗口 FirstActivity");
6.          Button bn = (Button)findViewById(R.id.bn);
7.              bn.setOnClickListener(new OnClickListener(){
8.                  public void onClick(View source)
9.                  {
10.                     //创建需要启动的 Activity 对应的 Intent
11.                     Intent intent = new
12.                     Intent(FirstActivity.this,SecondActivity.class);
13.                     //启动 intent 对应的 Activity
14.                     startActivity(intent);
15.                 }
16.             });
17.     }
18. ...
```

10) 打开 SecondActivity 源码，其相应代码如下：

```
1.  ...
2.  public void onCreate(Bundle savedInstanceState){
3.      super.onCreate(savedInstanceState);
4.      setContentView(R.layout.second);
5.      Button previous = (Button)findViewById(R.id.previous);
6.      previous.setOnClickListener(new OnClickListener(){
7.          public void onClick(View source)
8.          {
9.              //获取启动前 Activity 的上一个 Intent
10.             Intent intent = new
11.             Intent(SecondActivity.this,FirstActivity.class);
12.             //启动 intent 对应的 Activity
13.             startActivity(intent);
14.         }
15.     });
16.
17.     Button close = (Button)findViewById(R.id.Close);
18.     close.setOnClickListener(new OnClickListener(){
19.         public void onClick(View source)
20.         {
21.             //获取启动前 Activity 的上一个 Intent
22.             Intent intent = new
23.             Intent(SecondActivity.this,FirstActivity.class);
24.             //启动 intent 对应的 Activity
```

```
25.                 startActivity(intent);
26.                 finish();
27.             }
28.         });
29.     }
30. …
```

运行项目 chap05_2，可见如图 5-13 所示的界面。

单击图 5-13 中的 Button 按钮，可见如图 5-14 所示的运行界面。

图 5-13　chap05_2 运行界面（一）

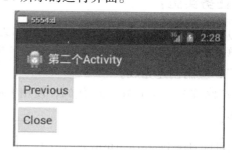

图 5-14　chap05_2 运行界面（二）

【知识学习】Activity 启动的方法

Acitivty 启动的方法有如下两种。

1）startActivity（Intent intent）：启动其他 Activity。如果关闭这个 Ativity，可以调用 finish 方法。

2）startActivityForResult（Intent intent，int requestCode）：以指定请求码 requestCode 的方式启动 Activity，且程序将会得到新启动 Activity 的结果，具体做法是，通过重写 onActivityResult 方法来获取。同样的，可以通过调用方法 finishActivity（int requestCode）来关闭这个 Activity。

读者可以看到，上面的两个方法都有一个 Intent 参数，它有什么含义呢？

前面的模块中也看到，在 AndroidMainfest.xml 中有一个 < intent – filter >，其作用是指定主入口 Activity 的。但一个 Android 应用有多个 Activity，当一个 Activity 需要启动另外一个 Activity 时，程序并没有直接告诉系统要启动哪个 Activity，而是通过 Intent 来表达自己的意图。一句话，Intent 封装了 Android 应用程序中需要启动某个组件的"意图"。要启动一个新的 Activity，开发者可以通过调用 Context 中的 startActivity，代码如下：

```
1. Intent intent = new Intent(this, ActivityDemo.class);
2. startActivity(intent);    //ActivityDemo 是需要启动的 Activity 类
```

上面代码的第 1 行中的 this 是指当前 Activity，ActivityDemo 是指需要启动的 Activity。

任务 3　Activity 组件生命周期的验证

【任务描述】

Android 应用程序可以是多线程的，并且在内存及处理器支持的情况下，Android 操作系统可以同时运行多个应用程序。应用程序可以在后台运行，但是在同一个时间内只能有一个活动的应用程序对用户可见。换句话说，在同一时间内，只能有一个应用程序的 Activity 处于前台。当一个 Activity 在 Android 应用中运行时，它的活动状态以 Activity 栈的形式进行管理。当前活动的 Activity 一定位于栈顶，随着不同应用的运行，每个 Activity 都可以从一种状态进入另一种状态。

为了让读者对 Activity 生命周期有更加深刻的理解，本任务将对生命周期进行验证，便于读者熟悉方法的调用次序。

【任务实施】

1）创建一个 Android 应用，项目名称为"chap05_3"，应用的名字为"LifeCycleTest"。连续单击 Next 按钮，输入 Activity 名字为"LifeCycleActivity"，输入布局文件名字为"main"，应用创建完成。

2）修改 LifeCycleActivity 代码，使用 Log 在 LogCat 窗口中输出相关信息，其相应程序如下所示。

```
1.   public class LifeCycleActivity extends Activity {
2.       private final static String TAG = "ActivityLifeCycleTest";
3.       protected void onCreate(Bundle savedInstanceState) {
4.           super.onCreate(savedInstanceState);
5.           setContentView(R.layout.main);
6.           Log.i(TAG,"调用 onCreate");
7.       }
8.       protected void onStart() {
9.           Log.i(TAG,"调用 onStart");
10.          super.onStart();
11.      }
12.      protected void onRestart() {
13.          Log.i(TAG,"调用 onRestart");
14.          super.onRestart();
15.      }
16.      protected void onResume() {
17.          Log.i(TAG,"调用 onResume");
18.          super.onResume();
19.      }
20.      protected void onPause() {
21.          Log.i(TAG,"调用 onPause");
22.          super.onPause();
23.      }
24.      protected void onStop() {
25.          Log.i(TAG,"调用 onStop");
26.          super.onStop();
```

```
27.     }
28.     protected void onDestroy( ){
29.         Log.i(TAG,"调用 onDestroy");
30.         super.onDestroy( );
31.     }
32. }
```

3）运行 APP 应用，打开 Logcat 视窗选择 Windows→Show View→LogCat 菜单命令，在打开应用时先后执行了 onCreate、onStart、onResume 这 3 个方法，如图 5-15 所示。

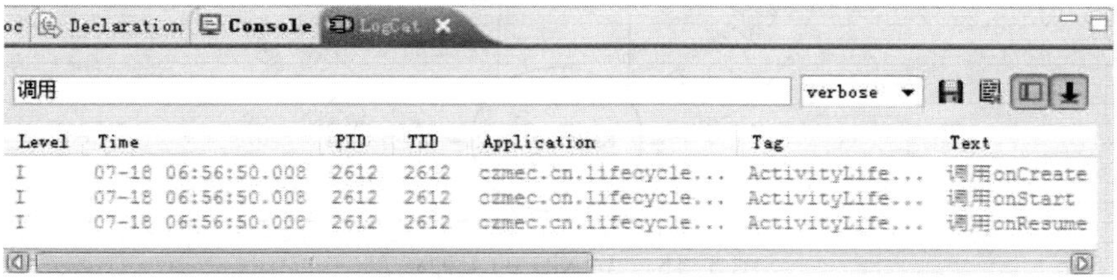

图 5-15 LogCat 窗口（一）

当按手机模拟器上的〈Back〉键时，这个应用将结束，这时候，开发者发现方法被调用的顺序是 onPause、onStop、onDestory，如图 5-16 所示。

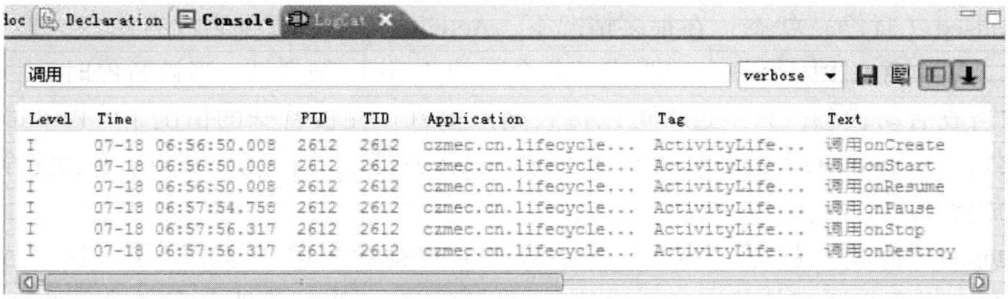

图 5-16 LogCat 窗口（二）

当 APP 应用运行后，当按〈Home〉键的时候，Activity 先后执行了 onPause、onStop 这两个方法，这时候应用程序并没有销毁，如图 5-17 所示。

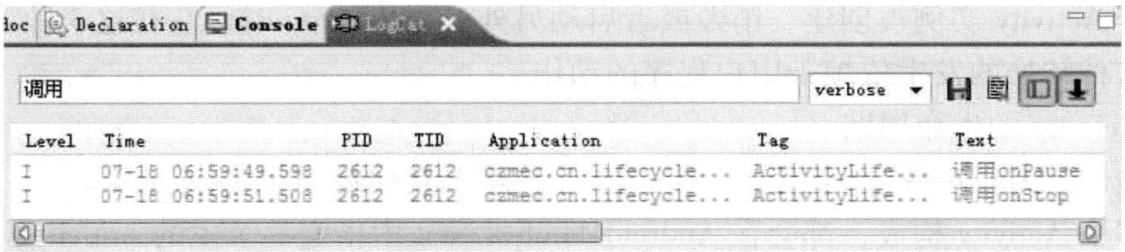

图 5-17 LogCat 窗口（三）

读者通过上面的案例可深刻理解了 Activity 的生命周期了。

【知识学习】Activity 生命周期

（1）Activity 的 4 种状态

归纳起来，Activity 有 4 种状态，如图 5-18 所示。

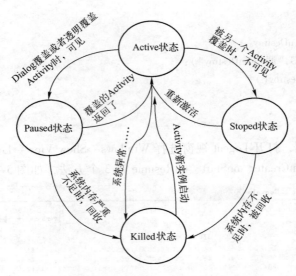

图 5-18　Activity 的 4 种状态

1) Active 状态：当一个 Activity 在栈顶时，它是可视的、有焦点的、可接受用户输入的。Android 试图尽最大可能保持它的活动状态，遏制其他 Activity 来确保当前活动 Activity 有足够的资源可使用。当另外一个 Activity 被激活时，这个将会被暂停。

2) Paused（暂停）状态：在很多情况下，Activity 可视，但是它没有焦点，换句话说它被暂停了。有可能的原因是一个透明或者非全屏的 Activity 被激活。当被暂停时，一个 Activity 仍会被当成活动状态，只不过不可以接收用户输入。在极特殊的情况下，Android 将会杀死一个暂停的 Activity 来为活动的 Activity 提供充足的资源。当一个 Activity 变为完全隐藏时，它将会变成停止。

3) Stoped（停止）状态：当一个 Activity 不是可视的，它"停止"了。这个 Activity 将仍然在内存中保存它所有的状态和会员信息。尽管如此，当其他地方需要内存时，它将是最有可能被释放资源的。当一个 Activity 停止后，一个很重要的步骤是要保存数据和当前 UI 状态。一旦一个 Activity 退出或关闭了，它将变为待用状态。

4) Killed（死亡）状态：在一个 Activity 被杀死后和被装载前，它是待用状态的。

当一个 Activity 实例被创建、销毁或者启动另外一个 Activity 时，它在这 4 种状态之间进行转换，这种转换的发生依赖于用户程序的动作。

(2) Activity 的生命周期

简单地说，Activity 就是布满整个窗口或者悬浮于其他窗口上的交互界面。一个应用程序通常由多个 Activity 构成，都会在 AndroidManifest.xml 中指定一个主的 Activity，如 < action-nandroid:name = " android. intent. action. MAIN" />。

当程序第一次运行时，用户就会看到这个 Activity，这个 Activity 可以通过启动其他的 Activity 进行相关操作。当启动其他的 Activity 时，当前的这个 Activity 将会停止，新的 Activity 将会压入栈中，同时获取用户焦点，这时就可在这个 Activity 上操作了。都知道栈遵循先进后出的原则，那么当用户按手机模拟器上的〈Back Space〉键时，当前的这个 Activity 销毁，前一个 Activity 重新恢复。Activity 的生命周期如图 5-19 所示。

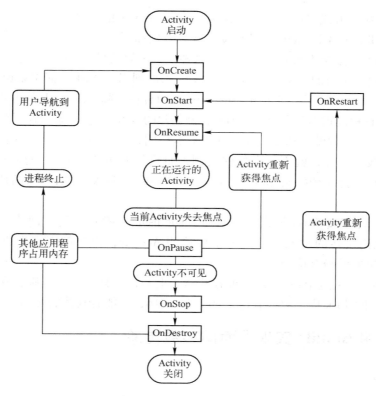

图 5-19 Activity 的生命周期

下面对 Activity 生命周期中几个重要的方法进行说明。

1）protected void onCreate（Bundle savedInstanceState）：这是 Activity 实例被启动时调用的第一个方法。一般情况下，开发者都常通过覆盖该方法作为应用程序的一个入口点，在这里做一些初始化数据、设置用户界面等工作。

2）protected void onStart：在 onCreate 方法之后被调用，或者在 Activity 从 Stoped 状态转换为 Active 状态时被调用。

3）protected void onResume：在 Activity 从 Paused 状态转换到 Active 状态时被调用。一般在这里做数据恢复工作。

4）protected void onPause：在 Activity 从 Active 状态转换到 Paused 状态时被调用。一般在这里保存 Activity 的状态信息。

5）protected void onStop：在 Activity 从 Active 状态转换到 Stoped 状态时被调用。

6）protected void onDestroy：在 Active 被结束时调用，它是被结束时调用的最后一个方法，在这里一般做些释放资源、清理内存等工作。

7）protected void onRestart：在 Activity 从 Stoped 状态转换为 Active 状态时被调用，不经常使用。

（3）Activity 的加载模式

在 Android 的多 Activity 开发中，Activity 之间的跳转有多种方式，有时是普通地生成一个新实例，有时希望跳转到原来某个 Activity 实例，而不是生成大量的重复的 Activity。加载模式便是决定以哪种方式启动一个跳转到原来某个 Activity 的实例。

在 Android 里，有 4 种 Activity 的启动模式，分别如下。

1）standard 模式：标准模式，每次通过这种模式来启动目标 Activity 时，Android 总会为目标 Activity 创建一个新的实例，并将该 Activity 添加到当前 Task（简单来说，一个 Task 就是用户体验上的一个"应用"，它将相关的 Activity 组合在一起，以 stack 的方式管理，这就是 Task）栈中——这种模式不会启动新的 Task，新创建的 Activity 会被添加到原有的 Task 中。

2）singleTop：这种模式和 standard 模式基本相同，有一点区别是，如果已经有一个实例位于 Activity 栈的顶部，就不产生新的实例，而只是调用 Activity 中的 newInstance 方法。如果不位于栈顶，则会产生一个新的实例，并将它加载到 Task 栈顶，此时的情况与 standard 模式完全相同。

3）singleTask：singleTask 模式和后面的 singleInstance 模式都是只创建一个实例的。在这种模式下，无论跳转的对象是不是位于栈顶的 Activity，程序都不会生成一个新的实例（当然前提是栈里面已经有这个实例）。

4）singleInstance：这个跟 singleTask 模式基本上一样，只有一个区别：在这个模式下的 Activity 实例所处的 Task 中，只能有这个 Activity 实例，不能有其他的实例。

模块二　使用 Bundle 实现界面间参数传递

【模块描述】

一个 APP 应用有多个 Activity，在多个 Activity 之间切换时需要传递参数，下面基于 Bundle 实现界面间数据交换。

知识点	技能点
➢ Intent 的创建及使用 ➢ 简单数据类型的传递与接收 ➢ 复杂数据类型的传递与接收 ➢ Bundle 对象的概念及使用	➢ 使用 Intent 进行界面跳转时通过 Bundle 对象载体传递数据的方法及注意事项 ➢ 使用 Intent 进行界面跳转时通过 Bundle 对象载体接收数据的方法及注意事项 ➢ 简单数据类型的传递及接收方法 ➢ 通过类封装的数据进行传递及接收的方法

任务　数据传递的具体实现

【任务描述】

APP 应用的界面切换有时需要传递参数，下面介绍基于 Bundle 实现界面跳转时的数据传递。

【任务实施】

1）创建一个 Android 应用，项目名称为"chap05_4"，应用的名字为"BundleTest"。连续单击 Next 按钮，输入 Activity 名字为"RegisterActivity"，输入布局文件名字为"register"，

应用创建完成。

2）创建一个实体类 Person，用来封装用户注册的信息，其相应代码如下：

```
1.  public class Person implements Serializable {
2.      private String _Name;
3.      private String _Passwd;
4.      private String _Gender;
5.      public String getName( )
6.      {
7.          return _Name;
8.      }
9.      public String getPass( )
10.     {
11.         return _Passwd;
12.     }
13.     public String getGender( )
14.     {
15.         return _Gender;
16.     }
17.     public Person(String Name, String Passwd, String Gender)
18.     {
19.         this._Name = Name;
20.         this._Passwd = Passwd;
21.         this._Gender = Gender;
22.     }
23. }
```

在定义 Person 类时必须实现接口 getSerializable，用以定义一个可序列化的 Person 类。

3）打开 register.xml 文件，按照如图 5-20 所示的内容进行修改。

4）同理，创建一个用来显示注册信息的 ShowActivity 及布局文件 show.xml。show.xml 布局文件如图 5-21 所示。

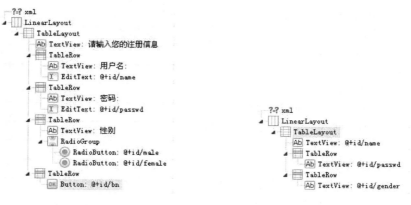

图 5-20　修改注册页面布局　　　　图 5-21　show.xml 布局文件

用以创建 ShowActivity 的相应程序内容如下：

```
1.      …
2.          super.onCreate(savedInstanceState);
3.          setContentView(R.layout.show);
4.          TextView name = (TextView)findViewById(R.id.name);
5.          TextView passwd = (TextView)findViewById(R.id.passwd);
6.          TextView gender = (TextView)findViewById(R.id.gender);
7.          //获取启动该 Result 的 Intent
8.          Intent intent = getIntent();
9.          //获取该 Intent 所携带的数据
10.         Bundle data = intent.getExtras();
11.         //从 Bundle 包中取出数据
12.         Person p = (Person)data.getSerializable("person");
13.         name.setText("用户名:" + p.getName());
14.         passwd.setText("密码:" + p.getPass());
15.         gender.setText("性别:" + p.getGender());
16.     …
```

5) 打开 RegisterActivity，其相应程序代码如下：

```
1.      …
2.          super.onCreate(savedInstanceState);
3.          setContentView(R.layout.register);
4.          this.setTitle("用户注册");
5.          Button bn = (Button)findViewById(R.id.bn);
6.          bn.setOnClickListener(new OnClickListener()
7.          {
8.              public void onClick(View v)
9.              {
10.                 EditText name = (EditText)findViewById(R.id.name);
11.                 EditText passwd = (EditText)findViewById(R.id.passwd);
12.                 RadioButton male = (RadioButton)findViewById(R.id.male);
13.                 String gender = male.isChecked()? "男" : "女";
14.                 Person p = new
                Person(name.getText().toString(),passwd.getText().toString(),gender);
15.                 //创建 Bundle 对象
16.                 Bundle data = new Bundle();
17.                 data.putSerializable("person",p);
18.                 //创建一个 Intent
19.                 Intent intent = new
                        Intent(RegisterActivity.this,ShowActivity.class);
20.                 intent.putExtras(data);
```

```
21.                    //启动 Intent 对应的 Activity
22.                    startActivity(intent);
23.                }
24.            });
25.    …
```

运行 chap05_4 应用,界面如图 5-22 所示。

单击"注册"按钮后的信息界面如图 5-23 所示。

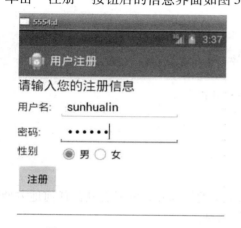

图 5-22 chap05_4 运行界面

图 5-23 注册信息界面

【知识学习】Activity 的几种跳转方式

(1) 显式调用方法

有如下 4 种显式调用方法。

1) 方法 1:Intent intent = new Intent(this,OtherActivity.class);

2) 方法 2:Intent intent2 = new Intent();
        intent2.setClass(this,OtherActivity.class);

3) 方法 3:intent2.setClassName(this,"com.zy.MutiActivity.OtherActivity");
此方式可用于打开其他应用。

4) 方法 4:intent2.setComponent(new ComponentName(this,OtherActivity.class));startActivity(intent2)。

(2) 隐式调用方法

只要 action、category、data 与要跳转到的 Activity 在 AndroidManifest.xml 中的设置匹配即可,如图 5-24 所示。

(3) 跳转到另一个 Activity 后需返回时能返回数据

在跳转的 Activity 端,调用 startActivityForResult(intent2,1) 后可跳转到下一个 Activity,其中,第一个参数为传入的意图对象,第二个为设置的请求码。跳转到第二个 Activity 后,调用 setResult(100,intent) 方法可返回上一个 Activity,其中,第一个参数为结果码,第二个为传入的意图对象。第一个 Activity 通过 onActivityResult 方法获得返回的数据。

例如,开发者要给"收件人"Activity 传递信息,则可通过下面这封 E – mail 将消息传

```xml
<activity android:name="com.zy.MutiActivity.OtherActivity">
    <intent-filter >
        <action android:name="com.zy.test.action"/>
        <category android:name="com.zy.java"/>
        <!-- 因为startActivity()方法中内置了该类别,索引必须加上此类别,否则android.intent.category.DEFAULT无法跳转-->
        <category android:name="android.intent.category.DEFAULT" />
        <!-- 设置了哪些,则哪些必须匹配,没设置的可以任意写 -->
        <data android:scheme="zhengyi" android:host="zy.com" android:path="/introduction"/>
        <!-- 数据类型 -->
        <data android:mimeType="txt/plain"/>
    </intent-filter>
</activity>
```

图 5-24 隐式调用

递出去,其相应程序代码如下:

1. Intent intent = new Intent(CurrentActivity.this,OtherActivity.class);
2. //创建一个带"收件人地址"的 E-mail
3. Bundle bundle = new Bundle();   //创建 E-mail 内容
4. bundle.putBoolean("boolean_key",true);//编写内容
5. bundle.putString("string_key","string_value");
6. intent.putExtra("key",bundle);//封装 E-mail
7. startActivity(intent);//启动新的 Activity

那么"收件人"该如何收信呢?在 OtherActivity 类的 onCreate 或者其他任何地方使用下面的代码就可以打开这封"E-mail"阅读其中的信息:

1. Intent intent = getIntent();//收取 E-mail
2. Bundle bundle = intent.getBundleExtra("key");//打开 E-mail
3. bundle.getBoolean("boolean_key");//读取内容
4. bundle.getString("string_key");

【问题研讨】 Activity 使用 Bundle 实现 Activity 间数据传递

前面介绍了如何启动一个新的 Activity 的方法,但如果要从当前的 Activity 中传递数据到新的 Activity,如何做呢?

在 Android 中,两个 Activity 之间是靠 Intent 传递信息的,可以使用如下方法。

1) putExtras(Bundle data):向 Intent 中放入需要传递的数据。
2) putXxx(String key,Xxx date):向 Bundle 放入 int、long 等各种类型的数据。
3) putSerializable(String key,Serializable date):向 Bundle 中放入一个可序列化的对象。
4) getXxx(String key):从 Bundle 取出 Int、Long 等各种数据类型的数据。
5) getSerializable(String Key,Serializable data):从 Bundle 取出一个可序列化的对象。

在当前的 Activity 中传递数据到新的 Activity 中,其相应代码如下:

1. Intent intent = new Intent(this,ActivityDemo.class);
2. Bundle bundle = new Bundle();
3. bundle.putBoolean("bool_key",true);
4. intent.putExtras(bundle);
5. startActivity(intent);

有时开发者需要启动带返回值的 Activity，简单地说就是需要新启动的 Activity 返回时将值传递给启动它的 Activity，其相应代码如下：

1. Intent intent = new Intent(ActivityLifeDemo. this, RevalueActivity. class);
2. startActivityForResult(intent, 0x1001);

其中，ActivityLifeDemo 是当前的 Activity，以启动 RevalueActivity，开发者在当前的 ActivityLifeDemo 中需要获取 RevalueActivity 传回来的值。那么，在 RevalueActivity 中就必须这样写：

1. Intent intent = new Intent( );
2. intent. putExtra("revalue_key","haha – revalueActivity");
3. setResult(0x1001, intent);

那么 revalue_key 值在哪里获取呢？必须重写 onActivityResult 方法，通过判断 requestCode 来确定：

1. if(requestCode == 0x1001) {
2.   String str = data. getStringExtra("revalue_key");
3.   Log. i(TAG,"返回的值为：" + str); }

模块三　Android 事件处理编程

【模块描述】

Android 应用程序中的事件处理秉承了 JavaSE 图形用户界面的处理方式和风格。

Android 平台的事件处理机制有两种。

1）基于监听的事件处理：对于 Android 基于监听的事件处理，主要的做法是为 Android 界面组件绑定特定的事件监听器。

2）基于回调的事件处理：对于 Android 基于回调的事件处理，主要的方法是重写 Android 组件特定的回调方法，或者重写 Activity 的回调方法。

项目三模块二任务 1 中实现的移动的小球实例就是采用回调方式进行事件处理实现的，本项目主要实现基于监听的事件处理机制的编程模式。

知识点	技能点
➢ Android 事件处理机制涉及的三要素 ➢ Android 事件处理机制 ➢ 基于监听的事件处理机制和基于回调的事件处理机制的优缺点 ➢ 不同事件监听机制下事件监听器的实现及各自优缺点	➢ 内部类实现事件监听器 ➢ 匿名内部类实现事件监听器 ➢ 外部类实现事件监听器 ➢ Activity 实现事件监听器 ➢ 为组件绑定相关属性实现事件监听

任务1 内部类实现事件监听器

【任务描述】

使用内部类实现事件监听器是指在 Activity 中定义一个事件监听器,这个事件监听器本身是一个内部类。在这个内部类中定义需要监听的事件。

【任务实施】

1)创建一个 Android 应用,项目名称为"chap05_5",应用的名称为"EventListenerTest"。连续单击 Next 按钮,输入 Activity 名字为"EventListenerActivity",输入"布局文件"名字为"main",应用创建完成。

2)打开 main 文件,修改后内容如图 5-25 所示。

图 5-25 修改 main 布局文件

3)打开 EventListenerActivity,其相应程序代码如下:

```
1.  public class EventListenerActivity extends Activity {
2.      protected void onCreate(Bundle savedInstanceState) {
3.          super.onCreate(savedInstanceState);
4.          setContentView(R.layout.main);
5.          Button btn = (Button)this.findViewById(R.id.btn);
6.          btn.setOnClickListener(new MyClickListener());
7.      }
8.      class MyClickListener implements View.OnClickListener
9.      {
10.         public void onClick(View v){
11.             TextView tv = (TextView)findViewById(R.id.txt);
12.             tv.setText("使用内部类实现的监听器哦");
13.         }
14.     }
15. }
```

上述代码第 8 行为按钮 btn 设置了事件监听器 MyClickListener,这个监听器在下面采用内部类进行定义。在定义时实现了 Android 的 View.OnClickListener 接口,并实现了 onClick 方法。运行 chap05_5 应用,可得如图 5-26 所示的界面。

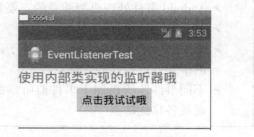

图 5-26 chap05_5 运行界面

任务 2　匿名内部类实现事件监听器

【任务描述】

上述案例介绍了采用内部类实现监听器的方法，但大部分时候，事件监听器只是临时使用一次，都没有什么实用价值，因此使用匿名内部类实现事件监听器更适合。

【任务实施】

1）实现思路是在页面中增加一个 Switch 开关，随着该状态开关的改变，界面布局中的线性布局的方向在水平布局和垂直布局之间切换。

2）创建一个 Android 应用，项目名称为"chap05_6"，应用的名称为"SwtichEvent-Test"。连续单击 Next 按钮，输入 Activity 名字为"MainActivity"，输入布局文件名字为"main"，应用创建完成。

3）打开 main 文件，其相应程序如下所示：

```
1.   <LinearLayout xmlns:android = "http://schemas.android.com/apk/res/android"
2.       android:layout_width = "fill_parent"
3.       android:layout_height = "wrap_content"
4.       android:orientation = "vertical" >
5.   <LinearLayout
6.       android:layout_width = "fill_parent"
7.       android:layout_height = "wrap_content"
8.       android:orientation = "vertical"
9.       android:gravity = "center" >
10.      <Switch
11.          android:id = "@ + id/sw"
12.          android:layout_width = "wrap_content"
13.          android:layout_height = "wrap_content"
14.          android:checked = "true"
15.          android:gravity = "center|start"
16.          android:textColorHighlight = "#ff0000"
17.          android:textOff = "纵向排列"
18.          android:textOn = "横向排列" />
19.   </LinearLayout>
20.   <LinearLayout
21.       android:layout_width = "fill_parent"
22.       android:layout_height = "wrap_content"
23.       android:orientation = "vertical"
24.       android:id = "@ + id/l"
25.       android:gravity = "center" >
26.       <TextView
```

```
27.        android:id = "@ + id/tv1"
28.        android:layout_width = "wrap_content"
29.        android:layout_height = "wrap_content"
30.        android:text = "测试文本一"
31.        android:textColor = "#ff0000"/>
32.    <TextView
33.        android:id = "@ + id/tv2"
34.        android:layout_width = "wrap_content"
35.        android:layout_height = "wrap_content"
36.        android:text = "测试文本二"
37.        android:textColor = "#ff0000"/>
38.    <TextView
39.        android:id = "@ + id/tv3"
40.        android:layout_width = "wrap_content"
41.        android:layout_height = "wrap_content"
42.        android:text = "测试文本三"
43.        android:textColor = "#ff0000"/>
44.    </LinearLayout>
45. </LinearLayout>
```

4) 打开 MainActivity,其相应程序代码如下:

```
1.  public class MainActivity extends Activity {
2.      protected void onCreate(Bundle savedInstanceState) {
3.          super.onCreate(savedInstanceState);
4.          setContentView(R.layout.main);
5.          Switch sw = (Switch)this.findViewById(R.id.sw);
6.          final LinearLayout l = (LinearLayout)this.findViewById(R.id.l);
7.          sw.setOnCheckedChangeListener(new OnCheckedChangeListener()
8.          {
9.              public void onCheckedChanged(CompoundButton buttonView,
10.                 boolean isChecked) {
11.                 if(isChecked)
12.                 {
13.                     l.setOrientation(1);
14.                 }
15.                 else
16.                 {
17.                     l.setOrientation(0);
18.                 }
19.             }
20.         });
```

21.   }
22.   }

上述代码的第 7 ～ 20 行即定义了一个匿名内部类，实现了 onCheckedChanged 方法。当开关的状态发生变化时，如从 on 切换到 off，将会调用这个方法。当开关处于 on 状态时，设置界面的线性布局为垂直布局（代码第 13 行）；当开关处于 off 状态时，设置界面的线性布局为水平方式（代码第 17 行）。开关为 on 时的运行界面如图 5-27 所示。

当拨动开关到 off 状态时，测试文本进行水平排列了，如图 5-28 所示。

图 5-27 开关为 on 时的运行界面 　　　　图 5-28 开关为 off 时的运行界面

任务 3 外部类实现事件监听器

【任务描述】

由于事件监听器通常隶属特定的 UI 界面，因此使用外部类定义事件监听器类很少见，主要是因为这样做不利于提高程序的内聚性。但如果某个事件监听器确实需要被多个 GUI 界面共享，则可以考虑使用外部类来实现。本任务主要使用外部类改写任务 2 中的 chap05_6 案例。

【任务实施】

1）创建一个监听器类 MyOnCheckedChangeListener，继承 OnCheckedChangeListener，其相应程序代码如下所示：

```
1.  public class MyOnCheckedChangeListener implements OnCheckedChangeListener {
2.      private Activity act;
3.      private Switch sw;
4.      private LinearLayout l ;
5.      public MyOnCheckedChangeListener( Activity act ,Switch sw,LinearLayout l)
6.      {
7.          this. act = act;
8.          this. sw = sw;
9.          this. l = l;
10.     }
11.     public void onCheckedChanged( CompoundButton buttView,boolean isChecked) {
```

```
12.              if(isChecked)
13.              {
14.                  l. setOrientation(1);
15.              }
16.              else
17.              {
18.                  l. setOrientation(0);
19.              }
20.          }
21.
22.     }
23. }
```

上述代码中,由于监听器类需要和界面中的组件绑定,因此首先定义了 Activity、Switch、LinearLayout 这 3 个组件,然后在第 5 行定义了构造法方法,用来实现对监听器类的初始化。于是监听器类实现了 OnCheckedChangeListener 接口,并实现了其中的 onCheckedChanged 方法。

2)修改 MainActivity,即删除第 7 ~ 20 行代码,增加如下代码。

```
1.  …
2.  MyOnCheckedChangeListener myListener = new
                 MyOnCheckedChangeListener(this,sw,l);
3.  sw. setOnCheckedChangeListener(myListener);
4.  …
```

任务4 Activity 实现事件监听器

【任务描述】

更简单地实现监听器的方法就是采用 Activity 本身。此时,这个 Activity 需要实现 OnClickListener 接口。本任务将在任务1 的项目5-5 的基础上进行改写。

【任务实施】

用 Activity 改写上例的程序代码如下。

```
1.  public class EventListenerActivity extends Activity implements OnClickListener{
2.      @Override
3.      protected void onCreate(Bundle savedInstanceState){
4.          super. onCreate(savedInstanceState);
5.          setContentView(R. layout. main);
6.          Button btn = (Button)this. findViewById(R. id. btn);
7.          TextView tv = (TextView)findViewById(R. id. txt);
```

```
8.          btn.setOnClickListener(this);
9.      }
10.     public void onClick(View v){
11.         TextView tv = (TextView)findViewById(R.id.txt);
12.         tv.setText("使用内部类实现的监听器");
13.     }
14. }
```

其中,onClick 方法用以重写了 OnClickListener 接口中的方法。

上述方法最大的优点就是简捷,但不可否认,这种做法的最大缺点就是:Activity 的主要职责是完成界面的初始化工作,但是此时又包含了事件处理方法,从而引起程序结构混乱。因此建议读者不要采用这种方法。

任务5 为组件绑定相关属性实现事件监听器

【任务描述】

Android 还有一种更加简单的实现事件监听器的方式,即直接在布局文件中为相关组件指定事件处理方法。本任务将在任务1的项目5-5的基础上进行改写。

【任务实施】

1)打开布局文件 main.xml,为按钮 btn 增加一个属性 onClick,指定这个属性的值为 MyClickHandler。

2)打开 MainActivity 文件,修改其中的 onCreate 方法,其相应程序内容如下:

```
1.  public class EventListenerActivity extends Activity {
2.      protected void onCreate(Bundle savedInstanceState) {
3.          super.onCreate(savedInstanceState);
4.          setContentView(R.layout.main);
5.          Button btn = (Button)this.findViewById(R.id.btn);
6.          Public void MyClickHandler(View v)
7.          {
8.              TextView tv = (TextView)findViewById(R.id.txt);
9.              Tv.setText("我是使用为组件绑定属性实现的");
10.         }
11.     }
12. }
```

任务6 Handler 消息传递编程

【任务描述】

当一个程序第一次启动时,Android 会同时启动一条主线程,主线程主要负责处理与 UI 相关的事件,如用户的按键事件、用户接触屏幕的事件、屏幕绘图事件,并把相关的事件分发到相应的组件进行处理，所以主线程通常又被称为 UI 线程。

出于性能优化考虑，Android 系统中的 UI 操作并不是线程安全的，如果多个线程并发地去操作同一个组件，可能导致线程安全问题。为了解决这一个问题，Android 有一条规则：只允许 UI 线程来修改 UI 组件的属性等，也就是说必须单线程模型，这样如果在 UI 界面进行一个耗时较长的数据更新等就会形成程序假死现象，也就是 ANR 异常，如果 20 s 中没有完成更新程序就会被强制关闭。

Android 的消息传递机制是另一种形式的"事件处理"，这种机制主要是为了解决 Android 应用的多线程问题——Android 平台不允许新启动的线程访问该 Activity 里的界面组件，这样就会导致新启动的线程无法改变新界面组件的属性值。但在实际 Android 应用开发中，尤其是涉及动画的游戏中，需要让新启动的线程周期地改变组件的属性值，这就需要借助 Handler 的消息传递机制来实现了。

本任务将基于 Handler 消息传递机制实现一个简单的图片自动播放器。

【任务实施】

1）创建一个 Android 应用，项目名称为 "chap05_7"，应用的名称为 "HandlerTest"。连续单击 Next 按钮，输入 Activity 名字为 "HandlerActivity"，输入布局文件名字为 "main"，在 main 中增加一个 id 为 photo 的 ImageView 组件，应用创建完成。

2）打开 HandlerActivity 文件，其程序内容如下。

```
1.    public class HandlerActivity extends Activity {
2.        //定义周期性显示的图片的 id
3.        int[ ] imageIds = new int[ ]
4.        {
5.            R. drawable. java,
6.            R. drawable. jsp,
7.            R. drawable. web,
8.            R. drawable. software
9.        };
10.       int currentImageId = 0;
11.       public void onCreate( Bundle savedInstanceState)
12.       {
13.           super. onCreate( savedInstanceState);
14.           setContentView( R. layout. main);
15.           final ImageView show = ( ImageView) findViewById( R. id. photo);
16.           final Handler myHandler = new Handler( )
17.           {
18.               //重写 Handler 的 handleMessage( Message msg)方法,该方法用于处理消息,
19.               //当新线程发送消息时,该方法会被自动调用
20.
21.               public void handleMessage( Message msg)
22.               {
```

```
23.                    //如果该消息是本程序所发送的
24.                    if( msg. what == 0x1233 )
25.                    {
26.                        //动态地修改所显示的图片
27.                        show. setImageResource( imageIds[ currentImageId ++ % imageIds. length ] );
28.                    }
29.                }
30.            };
31.            //定义一个计时器,让该计时器周期性地执行指定任务
32.            new Timer( ). schedule( new TimerTask( )
33.            {
34.                @Override
35.                public void run( )
36.                {
37.                    //新启动的线程无法访问该 Activity 里的组件
38.                    //所以需要通过 Handler 发送信息
39.                    Message msg = new Message( );
40.                    msg. what = 0x1233;
41.                    //发送消息,当执行 sendMessage 方法时,会让 Handler 的
42.                    //handleMessage( Message msg)方法自动被调用
42.                    myHandler. sendMessage( msg );
43.                }
44.            } ,0 ,800 );
45.        }
46.    }
```

代码的第 32 ~ 44 行通过 Timer 周期性地执行指定任务,Timer 对象可调度 IimerTask 对象,TimerTask 对象的作用就是启动一条新线程,由于 Android 不允许在新的线程中访问 Activity 里的界面组件,因此程序只能在新线程里发送一条消息,通知系统更新 ImageView 组件。第 16 ~ 30 行表示重写了 Handler 的 handlerMessage（Message msg）方法,该方法用于处理消息,即当新线程发送消息时,该方法会被自动回调,handlerMessage（Message msg）方法依然位于主线程,可以动态修改 ImageView 组件的属性。

3）将 4 张图片复制到 drawalbe - mdpi 目录中,运行 chap05_7,发现第一张图片可以运行显示出来,但是随之出现运行终止问题,如图 5-29 所示。

图 5-29　运行出错界面

上图运行错误是什么原因呢？此时打开 LogCat 窗口,可见到 "Out of memory on a 31961104 - byte allocation" 信息,是因为图片在加载到 Android 内存时因超出了内存的最大

限度而发生溢出现象，如图 5-30 所示。

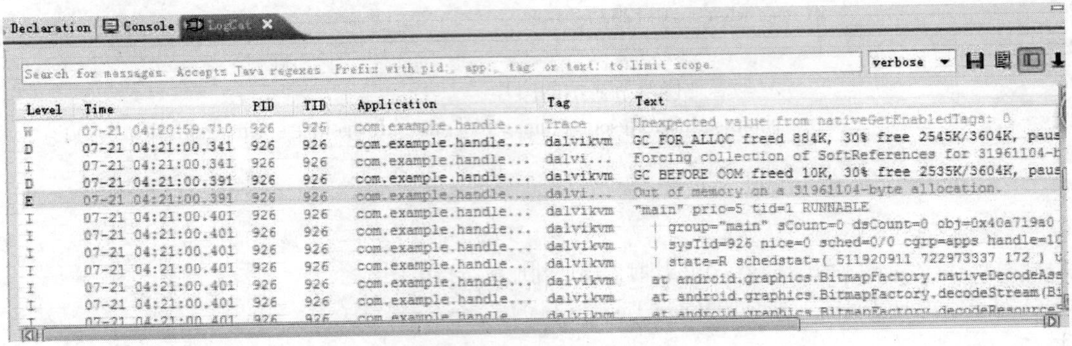

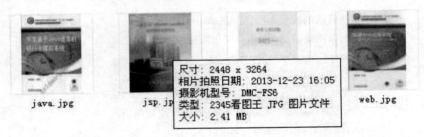

图 5-30　LogCat 出错信息

经查看，drawalbe - mdpi 目录中的 .jsp 图片的大小为 2.41 MB，如图 5-31 所示。

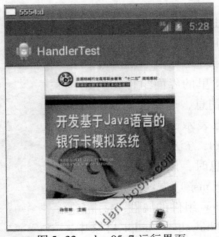

图 5-31　图片大小显示

将这个图片进行处理，修改约为 50 KB 后重新运行，可见如图 5-32 所示的运行界面，在这个界面中的图片每隔一段时间将会自动显示另一张图片。

图 5-32　chap05_7 运行界面

其实，上面的解决办法并不推荐，读者可以采用下面的方法，其相应程序代码如下所示：

```
1.   …
2.   public static Bitmap decodeSampledBitmapFromResource(Resources res,
3.           int resId,int reqWidth,int reqHeight){
4.       //第一次解析时将 inJustDecodeBounds 设置为 true,来获取图片大小
5.       final BitmapFactory.Options options = new BitmapFactory.Options();
6.       options.inJustDecodeBounds = true;
7.       BitmapFactory.decodeResource(res,resId,options);
8.       //调用上面定义的方法计算 inSampleSize 值
9.       options.inSampleSize = calculateInSampleSize(options,reqWidth,reqHeight);
10.      //使用获取到的 inSampleSize 值再次解析图片
11.      options.inJustDecodeBounds = false;
12.      return BitmapFactory.decodeResource(res,resId,options);
13.  }
14.
15.  public static int calculateInSampleSize(BitmapFactory.Options options,
16.          int reqWidth,int reqHeight){
17.      //源图片的高度和宽度
18.      final int height = options.outHeight;
19.      final int width = options.outWidth;
20.      int inSampleSize = 1;
21.      if(height > reqHeight || width > reqWidth){
22.      //计算出实际宽高和目标宽高的比率
23.      final int heightRatio = Math.round((float)height / (float)reqHeight);
24.      final int widthRatio = Math.round((float)width / (float)reqWidth);
25.  //选择宽和高中最小的比率作为 inSampleSize 的值,这样可以保证最终图片的宽和高
26.      //一定都会大于等于目标的宽和高
27.      inSampleSize = heightRatio < widthRatio ? heightRatio : widthRatio;
28.  }
29.      return inSampleSize;
30.  }
31.  …
32.  //下面代码表示简单地将任意一张图片压缩成 150 * 100 像素的缩略图,并在 ImageView 上展示
33.      show.setImageBitmap(decodeSampledBitmapFromResource(getResources(),
34.          imageIds[currentImageId ++ % imageIds.length],150,100));
35.  …
```

【知识学习】 Handler 类

Handler 类的主要作用有两个。

1) 在新启动的线程中发送消息。
2) 在主线程中获取、处理消息。

为了主线程能"适时"地处理新启动的线程所发送的消息,只能通过回调的方式来实

现,即开发者重写 Handler 类中处理消息的方法。当新启动的线程发送消息时,Hanlder 类中处理消息的方法被自动调用。

Handler 类包含如下方法用于发送、处理消息。

1) void handleMessage(Message msg):处理消息的方法,该方法经常被用于重写。

2) final boolean hasMessages(int what):检查消息队列中是否包含 what 属性为指定值的消息。

3) final boolean hasMessages(int what,Object object):检查消息队列中是否包含 what 属性为指定值、object 为指定对象的消息。

4) Message(int what):获取消息。

5) sendEmptyMessage(int what):发送空消息。

6) final boolean sendEmptyMessageDelayed(int what,long delayMills):指定多少毫秒后发送空消息。

7) final boolean sendMessage(Message msg):立即发送消息。

8) final boolean sendMessageDelayed(Message msg,long delayMills):指定多少毫秒后发送消息。

借助于上面这些方法,Android 应用程序可以方便地利用 Handler 来进行消息传递。

【任务拓展】

(1) 按钮 button 单击事件

button 单击事件相关的程序代码如下:

```
1.    button01.setOnClickListener(new Button.onClickListener()
2.    {
3.        public void onClick()
4.        {
5.            //编写按钮单击事件程序时需要调用的功能代码
6.        }
7.    });
```

以上程序中:

1) boolean onKeyMultiple(int keyCode,int repeatCount,KeyEvent event):用于在多个事件连续时发生按键重复,必须重载@Override 实现。

2) boolean onKeyDown(int keyCode,KeyEvent event):用于在按键按下时发生。

3) boolean onKeyUp(int keyCode,KeyEvent event):用于在按键进行释放时发生。

4) onTouchEvent(MotionEvent event):触摸屏事件,当在触摸屏上有动作时发生。

(2) checkbox 选择事件

checkbox 选择事件相关的程序代码如下。

```
1.    checkbox.setonCheckChangeListener(new CheckBox.onCheckChangedListener()
2.    {
3.        public void onCheckedChanged()
```

```
4.        {
5.            //编写按钮单击事件程序时需要调用的功能代码
6.        }
7.    });
```

（3）选择 item 事件使用方法

选择 item 事件使用方法的程序代码如下。

```
1. checkbox.setOnItemLongClickListener( new CheckBox.OnItemLongClickListener( )
2. {
3.        public void OnItemLongClick( )
4.        {
5.            //编写按钮单击事件时需要调用的功能代码
6.        }
7.    });
```

【练习】

对项目三中习题"模仿实现 QQ 登录布局界面"继续进行修改，要求如下：

1）为"登录"按钮增加一个事件监听器，用来监听"登录"事件；

2）为"记住密码"等复选框增加事件监听器，用来监听"被选中"事件；

3）当用户选中一些复选框并单击"登录"按钮后进入"用户信息"界面，并在界面中显示用户在登录界面中输入或选中的信息。

项目六 高级 UI 组件的应用

本项目主要介绍 AdapterView 及其子类 ListView、AutoCompleteTextView、GridView、ExpandableListView 等列表的用法。在介绍上述组件的同时，也重点讲解了 Adapter 适配器及其子类 ArrayAdapter、SimpleAdapter 等用法。最后详细地介绍了其他一些常用组件如 ProgressBar、DatePickerDialog、SearchView、TabHost 及 AlertDialog 等用法。

【知识目标】

- Adapter View 及其子类
- Adapter 及其子类
- ListView 的用法
- AutoCompleteTextView 的用法
- GridView 的用法
- ExpandableListView 的用法
- ArrayAdapter 的用法
- SimpleAdapter 的用法
- ProgressBar 的用法
- DatePickerDialog 的用法
- SearchView 的用法
- TabHost 的用法
- AlertDialog 的用法

【模块分解】

- 使用 ListView 及 ArrayAdapter、SimpleAdapter 实现列表
- 使用 AutoCompleteTextView 实现文本框自动补齐
- 使用 GridView 实现表格
- 使用 ExpandableListView 实现可扩展列表
- 使用 ProgressBar 实现进度条
- 使用 DatePickerDialog 实现日期选择
- 使用 TabHost 实现选项页
- 使用 AlertDialog 实现多种对话框

模块一 使用 ListView 显示列表数据

【模块描述】

ListView 是 Android 应用中使用非常频繁的组件，它能以垂直列表显示数据。用户可以

直接在 XML 文件中使用 ListView 组件，可以在定义 Activity 时继承 ListActivity，也可以使用 ArrayAdapter 数据适配器填充 ListView，还可以使用 SimpleAdapter 填充 ListView。本模块将详细讲解上述 4 种方式创建及填充 ListView 列表。

知识点	技能点
➢ Android MVC 框架 ➢ AdapterView 及其子类的特征 ➢ ListView 组件的属性及其用法 ➢ ListActivity 的使用场合 ➢ ArrayAdapter 数据适配器的用法 ➢ SimpleAdapter 数据适配器的用法	➢ 基于 XML 及 Java 编程的 ListView 两种创建方法 ➢ 使用 ListActivity 实现 ListView 的方法 ➢ 使用 ArrayAdapter 数据适配器填充 ListView 列表 ➢ 使用 SimpleAdapter 数据适配器填充 ListView 列表

任务 1　直接继承 ListActivity 创建 ListView

【任务描述】

本任务采用继承 ListActivity 的方式实现 ListView 列表。

【任务实施】

1）创建一个 Android 应用，设置名称为"chap06_1"，项目名称为"ListViewTest"，Activity 名称为"ListViewActivity"文件，布局文件名称为"main"。

2）打开 ListViewActivity，其相应程序内容如下：

```
1.  …
2.  protected void onCreate(Bundle savedInstanceState) {
3.      super.onCreate(savedInstanceState);
4.      String[] arr = {"小猫咪","小花狗","大黑猪","小兔子","老母鸡"};
5.      ArrayAdapter<String> adapter = new
6.  ArrayAdapter<String>(this,android.R.layout.simple_list_item_multiple_choice,arr);
7.      setListAdapter(adapter);
8.  }
9.  …
```

上述代码表示直接继承 ListActivity 实现了列表显示。需要注意的是，不用调用 setContentView 方法显示界面，而是直接调用 setListAdapter 方法将定义的适配器 Adapter 传入即可。这里主要是使用了 Adapter 适配器，开发者可以把它看成是 ListView 的数据源，ListView 要展示的数据都是以 Adapter 的形式传递给 ListView 的。这个 Adapter 很重要，Android 的用于传给集合控件（ListView、Spinner、GridView 等）的数据都是 Adapter 形式，这样的好处就是只要掌握了 Adapter，就可以很随意地给这些集合控件传递数据，因为它们使用的都是

Adapter。

运行 chap06_1 应用，得到的界面如图 6-1 所示。

上面的案例并没有使用在 main.xml 中定义一个 ListView 来布局 ListView，而是使用 ListActivity 中默认的 ListView 实现的。在 main.xml 中自定义一个 ListView，这样开发者可以很方便地控制 ListView 展示的布局、大小、背景色等属性。当然上个案例中，读者也可以不继承 ListActivity，直接继承 Activity，一样可以通过 findViewById 方法获取 ListView 后并使用它的方法来改变布局、大小和背景色等。

图 6-1 列表运行界面

【知识学习】AdapterView、Adapter 及 ListView

（1）AdapterView 与 Adapter

前面提到过一种经典的架构，即 MVC 框架，它是用数据模型 M（Model）存放数据，利用控制器 C（Controller）将数据显示在视图 V（View）上。Android 中有这样一种高级控件，其实现过程就类似于 MVC 框架。之所以称它高级，是因为它的使用不像其他控件一样拖动到界面上就能用，而是需要通过适配器将某些样式的数据或控件添加到其上而使用，这样的控件就是这里要详细介绍的 AdapterView。

AdapterView 组件是 Android 中重要的组件，它本身是一个抽象的类，其类图如图 6-2 所示。

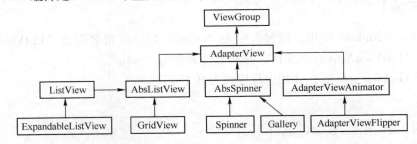

图 6-2 AdapterView 类图

AdapterView 继承了 ViewGroup，因此本质上是抽象的类，是一种容器。如果和 MVC 框架中的组件做类比的话，AdapterView 就是 View，是用来显示数据的。对于 AdapterView，一般编程时不直接使用，而是根据需求选择对应的其子类将数据进行列表显示。

1）AdapterView 及其子类有如下特征。

① 用来将数据进行列表显示。

② 将前端显示和后端数据分离。

③ 其内容不能通过 ListView.add 的形式添加列表项，需指定一个 Adapter 对象，通过它获得显示数据。

④ ListView 相当于 MVC 框架中的 V（视图），Adapter 相当于 MVC 框架中的 C（控制器），数据源相当于 MVC 框架中的 M（模型）。

⑤ 超出屏幕显示之后，自动加上滚动条。

2）从上图中可以看到，AdapterView 有 4 个重要的子类。

① ListView：表示列表，其中只能含有一个控件 TextView。
② Spinner：表示下拉列表，给用户提供选择。
③ Gallery：表示缩略图，已经被水平的 ScrollView 和 ViewPicker 取代，但也还算常用，是一个可以把子项进行中心锁定、水平滚动的列表。
④ GridView：表示网格图，以表格形式显示资源，可以左右滑动。

3）上述 AdapterView 的子类都有以下两个常用的事件。

① 用户单击列表项事件：为列表加载 setOnItemClickListener 监听，重写 onItemClick（发生单击事件的列表对象 ListView，被单击控件对象 view，在列表中的位置 position，被单击列表项的行 id）方法。

② 用户长按事件：为列表加载 setOnItemLongClickListener 监听，重写 onItemLongClick（发生单击事件的列表对象 ListView，被单击控件对象 view，在列表中的位置 position，被单击列表项的行 id）方法。

4）关于数据填充。

上述内容已经说明 AdapterView 是一个用来对数据进行列表显示的组件，相当于 MVC 中的 View。那么要显示的数据从何来呢？如何进行数据填充？这就需要数据适配器，即 Adapter 接口。Adapter 接口及实现类如图 6-3 所示。

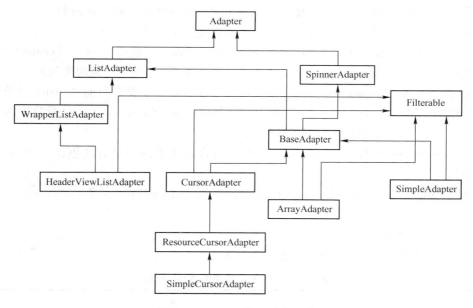

图 6-3 Adapter 接口及实现类

Adapter 就相当于 MVC 中的 Control 控制器，将数据源中的数据以某种样式（XML 文件）显示在视图中。下面对 Adapter 类的常用子类、通过继承 Adapter 类自定义的一个新 Adapter 类及 Adapter 进行数据填充与绑定内容的讲解。

① ArrayAdapter：比较简单易用，通常用于将数组或 List 集合中的数值进行列表显示，并且所处理的列表项内容必须是文本的情况。其数据源通常是数组或 List < String > 对象。

② SimpleAdapter：功能强大，可以将 List 集合中的多个对象列表显示，不仅可以处理列表项全是文本的情况，当列表项中还有其他控件时同样可以处理。其数据源只能为

List < Map < "键","值" >>形式的数据。

③ 自定义 Adapter：根据 XML 文件中定义的样式进行列表项的填充，适用性最强。而自定义 Adapter 的步骤如下。

  a. 创建类，继承自 BaseAdapter，对其重写的 4 个方法。
- Int getCount( )：返回的是数据源对象的个数，即列表项数。
- Object getItem（int position）：返回指定位置 position 上的列表。
- Long getItemId（int position）：返回指定位置处的行 id。
- View getView( )：返回列表项对应的视图。

  b. 实例化视图填充器。
  c. 用视图填充器根据 XML 文件实例化视图。
  d. 根据布局找到控件，并设置属性。
  e. 返回 View 视图。

④ SimpleCursorAdapter：与 SimpleAdapter 相似，专门用于把游标中的数据映像到列表中，它和数据库有关。

5）数据填充与绑定。

使用 Adapter 对 uI 组件进行数据填充与绑定的步骤如下：

① 声明 AdapterView 对象，可根据 id 利用 findViewById 方法找到此对象。

② 声明 Adapter 对象，可根据构造方法实例化此对象。具体如下。
- ArrayAdapter < 数据类型 > adapter = new ArrayAdapter < 数据类型 >（context：一般指当前 Activity 对象；layout：每个列表项显示的布局；data：数据源变量）。
- SimpleAdapter adapter = new SimpleAdapter（context：一般指当前 Activity 对象；data：数据源变量；layout：每个列表项显示的布局；new String[ ]{}：数据源中的"键"；new int[ ]{}：显示数据源的控件 id）。
- 如果是用户自定义的 Adapter 类，只要调用自定义类的构造方法创建一个 Adapter 对象即可。

③ 绑定 Adapter 对象到 Adapter 上的语法结构是：AdapterView 对象.setAdapter（Adapter 对象）。

（2）ListView 组件

ListView 组件常用的 XML 属性如表 6-1 所列。

表 6-1　ListView 常用的 XML 属性

属性名称	描述
android:choiceMode	使用的选择模式。默认状态下，list 没有选择模式。属性值有：none，值为 0，表示无选择模式；singleChoice，值为 1，表示最多可以有一项被选中；multipleChoice，值为 2，表示可以多项被选中
android:divider	分隔条，可以用颜色也可以用 drawable 分隔
android:dividerHeight	分隔符的高度。若没有指明高度，则用此分隔符固有的高度，必须为带单位的浮点数，如 14.5 sp。可用的单位有 px（pixel，像素）、dp（与密度无关的像素）、sp（基于字体大小的固定比例的像素），in（inches，英寸），mm（毫米）
android:entries	指定一个数组资源
android:footerDividersEnabled	设置成 flase 时，此 ListView 将不会在页脚视图前画分隔符。此属性默认值为 true。属性值必须设置为 true 或 false

(续)

属性名称	描 述
android:headerDividersEnabled	设置成 flase 时，此 ListView 将不会在页眉视图后画分隔符。此属性默认值为 true。属性值必须设置为 true 或 false
android:cacheColorHint	如果只换背景颜色，可指定 android:cacheColorHint 为所要的颜色，如用图片做背景，只要将 android:cacheColorHint 指定为透明（#00000000）
android:fadingEdge	设置上边和下边是否有黑色的阴影，值有 none、vertical 等
android:scrollbars	只有值为 horizontal｜vertical 的时候才会显示滚动条，并且会自动隐藏和显示；为 none 时则不会显示

任务 2　使用 XML 布局文件创建 ListView

【任务描述】

本任务采用 XML 布局文件的方式实现 ListView 列表。

【任务实施】

1）创建一个 Android 应用，名称为"chap06_2"，项目名称为"ListView2Test"，Activity 名称为"ListViewActivity"文件，布局文件名称为"main"。

2）打开 main，其相应程序内容如下：

```
1.    …
2.    <ListView
3.        android:id = "@+id/animal"
4.        android:layout_width = "match_parent"
5.        android:layout_height = "wrap_content"
6.        android:divider = "#f00"
7.        android:background = "#ccccff"  //银灰色
8.        android:dividerHeight = "2px"
9.        android:headerDividersEnabled = "false"
10.       android:entries = "@array/animal" >
11.   </ListView>
12.   …
```

3）在 layout/values 目录下创建 array.xml 资源文件，上述代码的第 10 行使用了 entries 属性指定了数组资源文件。将数组定义在 XML 文件中，其相应程序如下：

```
1.  <?xml version = "1.0" encoding = "utf-8"?>
2.  <resources>
3.      <string-array name = "animal">
4.          <item>小猫咪</item>
5.          <item>小花狗</item>
6.          <item>大黑猪</item>
7.          <item>小兔子</item>
```

8.   <item＞老母鸡＜/item＞
9. ＜/string – array＞
10. ＜/resources＞

运行 chap06_2 应用，可得如图 6-4 所示的界面。

任务 3 使用 ArrayAdapter 创建 ListView

【任务描述】

前面的任务分别是使用 ListActivity 和在布局文件中定义 ListView 的 entries 属性并在 arrays.xml 中指定数组资源实现的。下面直接使用 ArrayAdapter 创建 ListView 的方式对任务 2 进行修改。

图 6-4 chap06_2 运行界面

【任务实施】

1）创建一个 Android 应用，名称为"chap06_3"，项目名称为"ListView3Test"，Activity 名称为"List-ViewActivity"文件，布局文件名称为"main"。

2）打开 main，其相应程序内容如下：

1. ＜ListView
2.   android：id = "@ + id/animal"
3.   android：layout_width = "match_parent"
4.   android：layout_height = "wrap_content"
5.   android：divider = "#0f0"
6.   android：background = "#ccccff"
7.   android：dividerHeight = "2px"
8.   android：headerDividersEnabled = "false" ＞
9. ＜/ListView＞

3）在 layout 下创建自定义的列表格式文件 array_item，其相应程序内容如下：

1. ＜?xml version = "1.0" encoding = "utf – 8"?＞
2. ＜TextView xmlns：android = "http://schemas.android.com/apk/res/android"
3.   android：layout_width = "match_parent"
4.   android：layout_height = "wrap_content"
5.   android：id = "@ + id/tv"
6.   android：textSize = "18dp"
7.   android：shadowColor = "#f0f"
8.   android：shadowDx = "3"
9.   android：shadowDy = "3"
10.   android：shadowRadius = "1" ＞
11. ＜/TextView＞

4）打开 ListViewActivity，其相应程序内容如下：

```
1.  …
2.  protected void onCreate(Bundle savedInstanceState){
3.      super.onCreate(savedInstanceState);
4.      setContentView(R.layout.main);
5.      ListView lv = (ListView)this.findViewById(R.id.animal);
6.      String[] arr = {"小猫咪","小花狗","大黑猪","小兔子","老母鸡"};
7.      ArrayAdapter<String> adapter = new ArrayAdapter<String>(this, R.layout.array_item, arr);
8.      lv.setAdapter(adapter);
9.  }
10. …
```

上述第7行中的代码定义了一个 ArrayAdapter，其构造方法有3个参数，即分别是 Context、textViewResourceId 和数组（或 List 列表）。Context 代表访问整个 Android 应用的接口，几乎创建所有的组件都需要这个参数，这里用 this 代替；textViewResourceId 代表一个资源 id，该资源 id 代表一个 TextView，作为 ArrayAdapter 的列表项组件，这里采用 R.layout.array_item 资源文件；数组（或 List 列表）参数负责为列表项提供数据，这里使用 arr 数组。

运行 chap06_3，界面如图 6-5 所示。

图 6-5 chap06_3 运行界面

上面程序代码的第7行中的第3个参数是采用数组 arr，可以替换成 List 吗？回答是肯定的，开发者可以通过下面程序代码返回一个 List。

```
1.  …
2.  private List<String> getData(){
3.      List<String> data = new ArrayList<String>();
4.      data.add("小猫咪");
5.      data.add("小花狗");
6.      data.add("大黑猪");
7.      data.add("小兔子");
8.      data.add("老母鸡");
9.      return data;
10. }
11. …
```

请读者自行尝试一下吧。

任务4 使用 SimpleAdapter 创建 ListView

【任务描述】

上面的案例是使用 ArrayAdapter 实现列表显示，可以看出简单、易用，但是功能单一且有限,其列表中的每个列表项只能是 TextView,即只能是文本。如果列表选项比较复杂,比如要显示图片,则可以考虑使用 SimpleAdapter 组件。

【任务实施】

1)创建一个 Android 应用,名称为"chap06_4",项目名称为"SimpleAdapterTest",Activity 名称为"SimpleAdapterActivity",布局文件名称为"main"。

2)打开 main,id 为 animal 的 ListView,分别设置其宽度、高度为 fill_parent 和 wrap_content,ListView 中显示的数据由 SimpleAdapter 提供,其代码写在 Activity 中。

3)在 layout 目录下创建列表项的布局 simple_item.xml,其相应程序代码如下所示:

```
1.   <?xml version = "1.0" encoding = "utf-8"?>
2.   <LinearLayout xmlns:android = "http://schemas.android.com/apk/res/android"
3.       android:layout_width = "match_parent"
4.       android:layout_height = "wrap_content"
5.       android:orientation = "horizontal">
6.       <ImageView
7.           android:id = "@+id/photo"
8.           android:layout_width = "60dp"
9.           android:layout_height = "60dp"
10.          android:paddingLeft = "6dp" />
11.      <LinearLayout
12.          android:layout_width = "match_parent"
13.          android:layout_height = "wrap_content"
14.          android:orientation = "vertical">
15.          <TextView
16.              android:id = "@+id/name"
17.              android:layout_width = "wrap_content"
18.              android:layout_height = "wrap_content"
19.              android:paddingLeft = "6dp"
20.              android:textSize = "18dp"
21.              android:textColor = "#f0f"/>
22.          <TextView
23.              android:id = "@+id/description"
24.              android:layout_width = "wrap_content"
25.              android:layout_height = "wrap_content"
26.              android:paddingLeft = "6dp"
27.              android:textSize = "14dp"/>
28.      </LinearLayout>
29.  </LinearLayout>
```

上述程序表示布局文件对列表中的列表项的布局进行了定义。

4)打开 SimpleAdapterActivity,在其中定义 SimpleAdapter,其相应程序代码如下:

```
1.   public class SimpleAdapterActivity extends Activity
2.   {
```

```
3.      private String[ ] nameStrings = new String[ ]
4.                  {"小鸡","小狗","大猪","小猫","兔子"};
5.      private String[ ] description = {"我是一只小公鸡",
6.              "我是可爱的小狗狗",
7.              "我是一只大黑猪",
8.              "我是一只小猫咪",
9.              "我是只快乐的小兔子"};
10.     private int[ ] imageIds = new int[ ]
11.         {R. drawable. chick,R. drawable. dog,R. drawable. pig,R. drawable. cat,R. drawable. rabbit};
12.     @ Override
13.     protected void onCreate(Bundle savedInstanceState)
14.     {
15.         super. onCreate(savedInstanceState);
16.         setContentView(R. layout. main);
17.         List < Map < String,Object >> listItems = new ArrayList < Map < String,Object >> ();
18.         for(int i = 0;i < nameStrings. length;i ++ )
19.         {
20.             Map < String,Object > listItem = new HashMap < String,Object > ();
21.             listItem. put("photo",imageIds[i]);
22.             listItem. put("name",nameStrings[i]);
23.             listItem. put("description",description[i]);
24.             listItems. add(listItem);
25.         }
26.         SimpleAdapter simpleAdapter = new SimpleAdapter(this,listItems,
27.     R. layout. simple_item,new String[ ] {"name","photo","description"},new int[ ]
28.     {R. id. name,R. id. photo,R. id. description});
29.         ListView lView = (ListView)this. findViewById(R. id. animal);
30.         lView. setAdapter(simpleAdapter);
31.     }
32. }
```

上述程序代码的第 3 行和第 5 行分别定义了两个数组,用来分别存储列表名和相关描述;第 17 行定义了一个 List 集合,需要注意的是,这个集合中存储的对象是 Map 集合,而 Map 集合中的键 key 为 String 类型,值 value 为 Object 类型;第 18 ～ 25 行采用 for 循环对 List 进行填充数值。

第 26 行定义了一个 SimpleAdapter 对象,可以发现,定义这个对象需要 5 个参数:

① 第 1 个参数为 Context,使用 this;

② 第 2 个参数是一个 List < ? extends Map < String,? >> 类型的集合对象,该集合中每个 Map < String,? > 对象生成一个列表项。这里的这个参数是使用第 17 行定义的 listItems 集合;

③ 第三个参数(R. layout. simple_item,即第 3 步操作中在 layout 目录下创建的列表项布局 simple_item. xml 文件在 R 类中的映射)是一个 int 类型的整数,用于指定界面中的布局 id,即指定布局 simple_item. xml 文件作为列表项的布局组件;

④ 第4个参数是String[]数组类型，该参数决定了提取 Map 对象中的哪些 key 对应的 value 值来填充列表项。这就意味着这个数组中的值要和 Map 中定义的 key 一致；

⑤ 第5个参数是int[]类型的数组，该参数决定填充哪些组件。这就意味着这个数组中的值如 R. id. name、R. id. photo、R. id. description 即是 main. xml 中定义的组件，这个数组值的顺序要和第4个参数中定义的数组值的顺序一致。

当程序运行后，会发现列表中的列表有5项，其中第一个列表项组件中的数据是{name = " 小鸡 "，photo = " R. id. chick "，description = " 我是一只小公鸡 " }，当创建 SimpleAdapter 时，第5个参数及第4个参数用以指定使用 id 为 R. id. name 的组件显示 name 对应的值、使用 id 为 R. id. photo 的组件显示 photo 对应的值、使用 id 为 R. id. description 的组件显示 description 对应的值，这样第1个列表项组件所包含的3个组件都绑定了值。后面的4个列表项以此类推。

运行 chap06_4 应用，可得如图 6-6 所示的界面。

图 6-6　chap06_4 运行界面（一）

如果需要监听用户单击或选中了某个列表项的事件，可以通过 AdapterView 的 setIOnItemClickListener 方法为单击事件添加监听器，或通过 setOnItemSelectedListener 方法为列表项的选中事件添加监听器。例如，可以在上面的 Activity 中为 ListView 组件通过下面的程序代码内容绑定监听事件。

```
1.  …
2.  lView.setOnItemClickListener(new OnItemClickListener()
3.      {
4.          @Override
5.          public void onItemClick(AdapterView<?> parentView, View view, int position,
6.  long id)
7.          {
8.              String name = nameStrings[position];
9.              String desc = description[position];
10.             Animal animal = new Animal(name, desc);
11.             Bundle bundle = new Bundle();
12.             bundle.putSerializable("animal", animal);
13.             Intent intent = new
14.             Intent(SimpleAdapterActivity.this, SelectedActivity.class);
15.             intent.putExtras(bundle);
16.             startActivity(intent);
17.         }
18.     });
19. …
```

上述程序代码的第 10 行使用 Animal 类创建了一个 animal 对象，其中，Animal 类中有两个属性，即 name 和 description，它们对应 get、set 方法，还有一个带参数的构造方法。第 14 行定义了 Activity 跳转，即 SelectedActivity。其相应程序如下：

```
1.  …
2.  protected void onCreate(Bundle savedInstanceState)
3.  {
4.      super.onCreate(savedInstanceState);
5.      setContentView(R.layout.selected_main);
6.      this.setTitle("你选择的选项是");
7.      Intent intent = getIntent();
8.      Animal animal = (Animal)intent.getSerializableExtra("animal");
9.      TextView name = (TextView)findViewById(R.id.name);
10.     TextView description = (TextView)findViewById(R.id.description);
11.     name.setText(animal.getName());
12.     description.setText(animal.getDescription());
13. }
14. …
```

上述的代码程序第 5 行指定了跳转 Activity 的布局文件，即 selected_main，其相应程序如下：

```
1.  <?xml version = "1.0" encoding = "utf-8"?>
2.  <LinearLayout xmlns:android = "http://schemas.android.com/apk/res/android"
3.      android:layout_width = "match_parent"
4.      android:layout_height = "wrap_content"
5.      android:orientation = "horizontal" >
6.      <TextView
7.          android:id = "@+id/name"
8.          android:layout_width = "wrap_content"
9.          android:layout_height = "wrap_content"
10.         android:paddingLeft = "6dp"
11.         android:textSize = "30dp"
12.         android:textColor = "#f0f"/>
13.     <TextView
14.         android:id = "@+id/description"
15.         android:layout_width = "wrap_content"
16.         android:layout_height = "wrap_content"
17.         android:paddingLeft = "6dp"
18.         android:textSize = "25dp"/>
19. </LinearLayout>
```

当重新运行 chap06_4 后，界面如图 6-6 所示，当单击图中的第 4 行，即"小猫"列表项时，界面将会进入其详细页面，如图 6-7 所示。

图 6-7　chap06_4 运行界面（二）

模块二　文本框输入中自动提示列表的实现

【模块描述】

读者在使用百度搜索资料的时候，可以出现如图 6-8 所示的现象。

图 6-8　百度搜索

百度提示的词库并不是历史记录，是搜索引擎收集的当前热门搜索的内容。假如要在 Android 的应用实现如上功能，怎么做呢？本模块使用 AutoCompleteTextView 组件完成自动提示列表。

知识点	技能点
➢ Android 实现自动提示列表的方法 ➢ AutoCompleteTextView 的用法 ➢ MultiAutoCompleteTextView 的用法	基于 AutoCompleteTextView 实现自动提示列表

任务　基于 AutoCompleteTextView 实现自动提示列表

【任务描述】

基于 AutoCompleteTextView 实现自动提示列表，其提示框中的字符集可以使用一个 String 类型的数组填充。

【任务实施】

1）创建一个 Android 应用，名称为"chap06_5"，项目名称为"AutoCompleteTextView-Test"，Activity 名称为"AutoCompleteTextViewActivity"，布局文件为名称"main"。

2）打开 main，其相应程序代码如下：

```
1.  <?xml version="1.0" encoding="utf-8"?>
2.  <LinearLayout
3.      xmlns:android="http://schemas.android.com/apk/res/android"
4.      android:orientation="vertical"
5.      android:layout_width="fill_parent"
6.      android:layout_height="fill_parent">
7.      <TextView android:layout_width="fill_parent"
8.          android:layout_height="wrap_content"
9.          android:text="文本自动补齐"/>
10.     <LinearLayout android:layout_width="0px"
11.         android:layout_height="0px" android:focusable="true"
12.         android:focusableInTouchMode="true"></LinearLayout>
13.     <AutoCompleteTextView
14.         android:hint="请输入文字进行搜索"
15.         android:layout_height="wrap_content"
16.         android:layout_width="match_parent"
17.         android:id="@+id/txtView"
18.         android:textSize="18dp"
19.         android:completionThreshold="1">
20.     </AutoCompleteTextView>
21.     <Button android:text="搜索" android:id="@+id/search"
22.         android:layout_width="wrap_content"
23.         android:layout_height="wrap_content"></Button>
24. </LinearLayout>
```

3）打开 AutoCompleteTextViewActivity，其相应程序代码如下：

```
1.  public class AutoCompleteTextViewActivity extends Activity
2.  {
3.      AutoCompleteTextView autoCompleteTextView;
4.      String[] paramStrings = new String[]
5.      {"java面向对象编程",
6.       "java程序设计",
7.       "jsp案例教程",
8.       "jsp_servlet_javabean",
9.       "jiangsusheng",
10.      "jiangsu",
11.      "jianguo",
```

147

```
12.            "jaba",
13.            "jerry"};
14.
15.      protected void onCreate(Bundle savedInstanceState)
16.      {
17.            super.onCreate(savedInstanceState);
18.            setContentView(R.layout.main);
19.            ArrayAdapter<String> arrayAdapter = new ArrayAdapter<String>(this,
20.   android.R.layout.simple_dropdown_item_1line, paramStrings);
21.            autoCompleteTextView = (AutoCompleteTextView)findViewById(R.id.txtView);
22.            autoCompleteTextView.setAdapter(arrayAdapter);
23.
24.      }
25. }
```

运行 chap06_5 应用，界面如图 6-9 所示。

当在上述界面中输入"java"时，提示信息如图 6-10 所示。

图 6-9 chap06_5 运行界面（一）　　　图 6-10 chap06_5 运行界面（二）

【知识学习】 智能输入中自动提示列表

Android 中提供了两种智能输入框，它们是 AutoCompleteTextView、MultiAutoComplete-TextView。前者继承自 EditView 的可编辑的文本视图，能够实现动态匹配输入的内容，后者继承自 AutoCompleteTextView 的可编辑的文本视图，功能大致相同。

AutoCompleteTextView 是一个可编辑的文本视图。当用户输入时会显示一个下拉菜单提示信息，即建议性列表，用户可以从中选择一项，以完成输入。建议列表是从一个数据适配器获取的数据。它有 3 个重要的方法。

1）clearListSelection：清除选中的列表项。

2）dismissDropDown：如果已存在列表项则关闭下拉菜单。

3）Adapter：获取适配器。

其他一些方法的说明如下。

1）setDropDownHeight：用来设置提示下拉框的高度。注意，这只是限制了提示性下拉框的高度，提示数据集的个数并没有变化。

2）setThreshold：设置从输入第几个字符起出现提示。

3）setCompletionHint：设置提示框最下面所显示的文字。

4）setOnFocusChangeListener：里面包含 OnFocusChangeListener 监听器，设置焦点并触发事件。

5）showdropdown：让下拉列表框弹出来。

6）clearListSelection：去除 selector 样式，但只是暂时地去除，当用户再输入时又重新出现。

7）dismissDropDown：关闭下拉列表框。

8）enoughToFilter：这是一个是否满足过滤条件的方法，SDK 建议用户可以重写这个方法。

9）getAdapter：得到一个可过滤的列表适配器。

10）getDropDownAnchor：设置下拉菜单的定位"锚点"组件，如果没有指定该属性，将使用该 TextView 本身作为定位"锚点"组件。

11）getDropDownBackground：得到下拉列表框的背景色。

12）setDropDownBackgroundDrawable：设置下拉列表框的背景色。

13）setDropDownBackgroundResource：设置下拉列表框的背景资源。

14）setDropDownVerticalOffset：设置下拉列表框的垂直偏移量，即 List 里包含的数据项数目。

15）getDropDownVerticalOffset：得到下拉列表框的垂直偏移量。

16）setDropDownHorizontalOffset，设置水平偏移量。

17）setDropDownAnimationStyle：设置下拉列表框的弹出动画。

18）getThreshold：得到过滤字符个数。

19）setOnItemClickListener：设置下拉列表框事件。

20）getListSelection：得到下拉列表框选中的位置。

21）getOnItemClickListener：得到单项单击事件。

22）getOnItemSelectedListener：得到单项选中事件。

【任务拓展】

（1）自动提示框问题

1）上述提示框字符集是事先写好并存放在一个 String[ ] 数组的，但真正应用时不是这样的，是将用户输入的信息作为历史记录保存下来，当用户下次输入的时候，会自动到用户的输入历史记录中检索。请读者认真思考下，该如何实现？

2）提示：可以将历史记录存储在 sharepreference 中，参考程序代码如下。

```
1.  public class AutoCompleteTextViewActivity extends Activity implements
2.          OnClickListener
3.  {
4.      private AutoCompleteTextView autoTv;
5.      public void onCreate( Bundle savedInstanceState)
6.      {
```

```
7.        super.onCreate(savedInstanceState);
8.        setContentView(R.layout.main);
9.        autoTv = (AutoCompleteTextView)findViewById(R.id.autoCompleteTextView1);
10.       initAutoComplete("history",autoTv);
11.       Button search = (Button)findViewById(R.id.button1);
12.       search.setOnClickListener(this);
13.   }
14.
15.   @Override
16.   public void onClick(View v)
17.   {
18.       //这里可以设定:当搜索成功时,才执行保存操作
19.       saveHistory("history",autoTv);
20.   }
21.
22.   /**
23.    * 初始化 AutoCompleteTextView,最多显示 5 项提示,使 AutoCompleteTextView 在一开始获得
24. 焦点时自动提示
25.    * @param field
26.    * 保存在 sharedPreference 中的字段名
27.    * @param auto
28.    * 要操作的 AutoCompleteTextView
29.    */
30.   private void initAutoComplete(String field,AutoCompleteTextView auto)
31.   {
32.       SharedPreferences sp = getSharedPreferences("network_url",0);
33.       String longhistory = sp.getString("history","nothing");
34.       String[] hisArrays = longhistory.split(",");
35.       ArrayAdapter<String> adapter = new ArrayAdapter<String>(this,
36.               android.R.layout.simple_dropdown_item_1line,hisArrays);
37.       //只保留最近的 50 条记录
38.       if(hisArrays.length > 50)
39.       {
40.           String[] newArrays = new String[50];
41.           System.arraycopy(hisArrays,0,newArrays,0,50);
42.           adapter = new ArrayAdapter<String>(this,
43.                   android.R.layout.simple_dropdown_item_1line,newArrays);
44.       }
45.       auto.setAdapter(adapter);
46.       auto.setDropDownHeight(350);
47.       auto.setThreshold(1);
48.       auto.setCompletionHint("最近的 5 条记录");
```

```
49.         auto.setOnFocusChangeListener( new OnFocusChangeListener( )
50.         {
51.             public void onFocusChange( View v,boolean hasFocus)
52.             {
53.                 AutoCompleteTextView view = ( AutoCompleteTextView) v;
54.                 if( hasFocus)
55.                 {
56.                     view.showDropDown( );
57.                 }
58.             }
59.         });
60.     }
61.     /**
62.      * 把指定 AutoCompleteTextView 中的内容保存到 sharedPreference 中指定的字符段
63.      * @param field
64.      * 保存在 sharedPreference 中的字段名
65.      * @param auto
66.      * 要操作的 AutoCompleteTextView
67.      */
68.     private void saveHistory( String field,AutoCompleteTextView auto)
69.     {
70.         String text = auto.getText( ).toString( );
71.         SharedPreferences sp = getSharedPreferences( "network_url",0);
72.         String longhistory = sp.getString( field,"nothing" );
73.         if( !longhistory.contains( text + "," ))
74.         {
75.             StringBuilder sb = new StringBuilder( longhistory);
76.             sb.insert( 0,text + ",");
77.             sp.edit( ).putString( "history",sb.toString( )).commit( );
78.         }
79.     }
80. }
```

请读者自行试试吧。具体代码可参见随书资源的 Chap06_6 文件夹（在"第一篇"文件夹的"项目六"里）。

（2）进度条 ProgressBar 的用法

ProgressBar 是 Android 中重要的进度条组件，它可以动态地显示进度，避免长时间执行某个耗时的操作，避免让用户感觉程序失去了响应，从而更好地提高用户的体验感，非常实用。

下面看看比较经典的几种进度条的风格。

1）普通圆形 ProgressBar 如图 6-11、图 6-12 所示。

图 6-11 普通圆形进度条（一）　　　图 6-12 普通圆形进度条（二）

该类型进度条也就是一个表示运转的过程，例如发送短信、连接网络等，表示一个过程正在执行中。一般只要在 XML 布局中定义就可以实现；其程序代码如下：

```
1.  < progressBar android:id = "@ + id/widget43"
2.          android:layout_width = "wrap_content"
3.          android:layout_height = "wrap_content"
4.          android:layout_gravity = "center_vertical" >
5.  </ProgressBar >
```

2）超大号圆形 ProgressBar 如图 6-13 所示。

此时为其设置一个 style 风格属性后，该 ProgressBar 就有了一个风格，这里的超大号圆形 ProgressBar 的风格是 style = "? android:attr/progressBarStyleLarge"。

图 6-13 超大号圆形进度条

完整的 XML 定义的程序是：

```
1.  < progressBar android:id = "@ + id/widget196"
2.          android:layout_width = "wrap_content"
3.          android:layout_height = "wrap_content"
4.          style = "? android:attr/progressBarStyleLarge" >
5.  </ProgressBar >
```

3）小号圆形 ProgressBar 如图 6-14 所示。

图 6-14 小号圆形进度条

小号圆形 ProgressBar 对应的风格是 style = "? android:attr/progressBarStyleSmall"。
完整的 XML 定义的程序是：

```
1.  < progressBar android:id = "@ + id/widget108"
2.          android:layout_width = "wrap_content"
3.          android:layout_height = "wrap_content"
4.          style = "? android:attr/progressBarStyleSmall" >
5.  </ProgressBar >
```

4）标题型圆形 ProgressBar 如图 6-15、图 6-16 所示。

 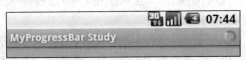

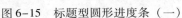

图 6-15 标题型圆形进度条（一）　　　图 6-16 标题型圆形进度条（二）

标题型圆形 ProgressBar 对应的风格是 style = "? android:attr/progressBarStyleSmallTitle" 完整的 XML 定义的程序是：

```
1.    <progressBar android:id = "@+id/widget110"
2.        android:layout_width = "wrap_content"
3.        android:layout_height = "wrap_content"
4.        style = "? android:attr/progressBarStyleSmallTitle" >
5.    </ProgressBar>
```

实现标题型圆形进度条的程序代码内容如下：

```
1.    protected void onCreate(Bundle savedInstanceState) {
2.        super.onCreate(savedInstanceState);
3.        requestWindowFeature(Window.FEATURE_INDETERMINATE_PROGRESS);
4.        //请求窗口特色风格,这里设置成风格不明确的进度条
5.        setContentView(R.layout.second);
6.        setProgressBarIndeterminateVisibility(true);
7.        //设置标题栏中不明确的进度条是否可以显示
8.    }
```

5）长条形进度条如图 6-17 所示。

图 6-17　长条形进度条

① 首先在 XML 进行布局，其程序为

```
1.    <progressBar android:id = "@+id/progressbar_updown"
2.        android:layout_width = "200dp"
3.        android:layout_height = "wrap_content"
4.        style = "? android:attr/progressBarStyleHorizontal"
5.        android:layout_gravity = "center_vertical"
6.        android:max = "100"
7.        android:progress = "50"
8.        android:secondaryProgress = "70"    >
9.    </progressBar>
```

说明：

```
style = "? android:attr/progressBarStyleHorizontal"      //设置风格为长条形
android:max = "100"                                      //最大进度值为 100
android:progress = "50"                                  //初始化的进度值
android:secondaryProgress = "70"                         //初始化的底层第二个进度值
android:layout_gravity = "center_vertical"               //垂直居中
```

② 实现长条形进度条的程序代码内容如下：

```
1.   private ProgressBar myProgressBar;
2.   //定义 ProgressBar
3.   myProgressBar = (ProgressBar)findViewById(R.id.progressbar_updown);
4.   //ProgressBar 通过 id 来从 XML 中获取
5.   myProgressBar.incrementProgressBy(5);
6.   //ProgressBar 进度值增加 5
7.   myProgressBar.incrementProgressBy(-5);
8.   //ProgressBar 进度值减少 5
9.   myProgressBar.incrementSecondaryProgressBy(5);
10.  //ProgressBar 背后的第二个进度条,进度值增加 5
11.  myProgressBar.incrementSecondaryProgressBy(-5);
12.  //ProgressBar 背后的第二个进度条,进度值减少 5
```

6）页面标题中的长条形进度条如图 6-18 所示。

实现该长条形进度条的程序代码实现如下。

图 6-18　页面标题中的长条形进度条

```
1.   requestWindowFeature(Window.FEATURE_PROGRESS);
2.   //请求一个窗口特色风格
3.   setContentView(R.layout.main);
4.   setProgressBarVisibility(true);
5.   //设置进度条可视
6.   setProgress(myProgressBar.getProgress()*100);
7.   //设置标题栏中前面的一个进度条的进度值
8.   setSecondaryProgress(myProgressBar.getSecondaryProgress()*100);
9.   //设置标题栏中后面的一个进度条的进度值
10.  //ProgressBar.getSecondaryProgress 用来获取其他进度条的进度值
```

7）ProgressDialog 中的圆形进度条如图 6-19 所示。

图 6-19　ProgressDialog 中的圆形进度条

ProgressDialog 一般用来表示一个系统任务或是开启任务时候的进度,有一种稍等的意思,其相应程序代码实现如下:

1.　ProgressDialog mypDialog = new ProgressDialog(this);
2.　　mypDialog.setProgressStyle(ProgressDialog.STYLE_SPINNER);//实例化
3.　　//设置进度条风格,风格为圆形、旋转的
4.　　mypDialog.setTitle("Google");
5.　　//设置 ProgressDialog 标题
6.　　mypDialog.setMessage(getResources().getString(R.string.second));
7.　　//设置 ProgressDialog 提示信息
8.　　mypDialog.setIcon(R.drawable.android);
9.　　//设置 ProgressDialog 标题图标
10.　mypDialog.setButton("Google",this);
11.　//设置 ProgressDialog 的一个 Button
12.　mypDialog.setIndeterminate(false);
13.　//设置 ProgressDialog 的进度条是否明确
14.　mypDialog.setCancelable(true);
15.　//设置 ProgressDialog 是否可以按〈Backspace〉键取消
16.　mypDialog.show();
17.　//让 ProgressDialog 显示

8) ProgressDialog 中的长条形进度条如图 6-20 所示。

图 6-20　ProgressDialog 中的长条形进度条

其相应程序代码实现如下:

1.　ProgressDialog mypDialog = new ProgressDialog(this);
2.　　//实例化
3.　　mypDialog.setProgressStyle(ProgressDialog.STYLE_HORIZONTAL);
4.　　//设置进度条风格,风格为长条形、有刻度的
5.　　mypDialog.setTitle("地狱怒兽");
6.　　//设置 ProgressDialog 标题
7.　　mypDialog.setMessage(getResources().getString(R.string.second));
8.　　//设置 ProgressDialog 提示信息
9.　　mypDialog.setIcon(R.drawable.android);

```
10.    //设置 ProgressDialog 标题图标
11.    mypDialog.setProgress(59);
12.    //设置 ProgressDialog 进度条进度
13.    mypDialog.setButton("地狱曙光",this);
14.    //设置 ProgressDialog 的一个 Button
15.    mypDialog.setIndeterminate(false);
16.    //设置 ProgressDialog 的进度条是否明确
17.    mypDialog.setCancelable(true);
18.    //设置 ProgressDialog 是否可以按〈Backspace〉键取消
19.    mypDialog.show();
20.    //让 ProgressDialog 显示
```

模块三　图片浏览器的实现

【模块描述】

GridView 和 AdapterViewFlipper 组件可以用来实现带预览功能的图片浏览器和图片自动播放器。本模块将基于这两个组件实现图片浏览器和播放器。

知识点	技能点
➢ GridView 常用的 XML 属性和方法 ➢ GridView 和 ListView 的区别 ➢ GridView 的事件监听和数据绑定 ➢ AdapterViewFlipper 的常用属性和用法	➢ 使用 GridView 组件实现带预览功能的图片浏览器 ➢ 使用 AdapterViewFlipper 组件实现图片自动播放器

任务1　用 GridView 实现带预览功能的图片浏览器

【任务描述】

使用 GridView 组件实现带预览功能的图片浏览器。

【任务实施】

1) 创建一个 Android 应用,名称为"chap06_7",项目名称为"GridViewTest",Activity 名称为"GridViewActivity"文件,布局文件名称为"main"。

2) 打开 main,其相应程序代码如下:

```
1.    <LinearLayout xmlns:android="http://schemas.android.com/apk/res/android"
2.        android:layout_width="fill_parent"
3.        android:layout_height="fill_parent"
4.        android:orientation="vertical"
5.        android:gravity="center_horizontal" >
6.
```

```
7.    <GridView
8.        android:id = "@+id/gridview"
9.        android:layout_width = "wrap_content"
10.       android:layout_height = "200dp"
11.       android:numColumns = "auto_fit"
12.       android:verticalSpacing = "10dp"
13.       android:horizontalSpacing = "10dp"
14.       android:columnWidth = "90dp"
15.       android:stretchMode = "columnWidth"
16.       android:gravity = "center"/>
17.   <LinearLayout
18.       android:layout_width = "220dp"
19.       android:layout_height = "170dp"
20.       android:orientation = "vertical"
21.       android:gravity = "center_horizontal">
22.       <ImageView
23.           android:id = "@+id/imageView"
24.           android:layout_width = "wrap_content"
25.           android:layout_height = "wrap_content"/>
26.   </LinearLayout>
27. </LinearLayout>
```

3）创建一个独立的 ImageAdapter 类，继承 BaseAdapter，重写其中的一些方法（例如下面的第 22、28、35、40 行代码）其相应程序代码如下所示：

```
1.  public class ImageAdapter extends BaseAdapter
2.  {
3.      //定义 Context
4.      private Context     mContext;
5.      //定义整型数组,即图片源
6.      public Integer[]  mImageIds =
7.      {
8.              R.drawable.img1,
9.              R.drawable.img2,
10.             R.drawable.img3,
11.             R.drawable.img4,
12.             R.drawable.img5,
13.             R.drawable.img6
14.     };
15.
16.     public ImageAdapter(Context c)
17.     {
```

```
18.         mContext = c;
19.     }
20.
21.     //获取图片的个数
22.     public int getCount( )
23.     {
24.         return mImageIds.length;
25.     }
26.
27.     //获取图片在库中的位置
28.     public Object getItem( int position)
29.     {
30.         return position;
31.     }
32.
33.
34.     //获取图片 id
35.     public long getItemId( int position)
36.     {
37.         return position;
38.     }
39.
40.     public View getView( int position, View convertView, ViewGroup parent)
41.     {
42.         ImageView imageView;
43.         if( convertView == null)
44.         {
45.             //给 ImageView 设置资源
46.             imageView = new ImageView( mContext);
47.             //设置布局,图片以 120×120 显示
48.             imageView.setLayoutParams( new GridView.LayoutParams(85,85));
49.             //设置显示比例类型
50.             imageView.setScaleType( ImageView.ScaleType.FIT_CENTER);
51.         }
52.         else
53.         {
54.             imageView = ( ImageView)convertView;
55.         }
56.         imageView.setImageResource( mImageIds[ position]);
57.         return imageView;
58.     }
59. }
```

4）打开 GridViewActivity，其相应程序代码如下：

```
1.  public class GridViewActivity extends Activity
2.  {
3.      protected void onCreate(Bundle savedInstanceState)
4.      {
5.          super.onCreate(savedInstanceState);
6.          setContentView(R.layout.main);
7.          //取得 GridView 对象
8.          GridView gridview = (GridView)findViewById(R.id.gridview);
9.          final ImageView imageView = (ImageView)findViewById(R.id.imageView);
10.         //添加元素给 gridview
11.         final  ImageAdapter imageAdapter =  new ImageAdapter(this);
12.         gridview.setAdapter(imageAdapter);
13.         //事件监听
14.         gridview.setOnItemClickListener(new OnItemClickListener(){
15.             public void onItemClick(AdapterView<?> parent, View v, int position, long id)
16.             {
17.                 Toast.makeText(GridViewActivity.this,"你选择了" + (position+1) + ""
18.                 号图片", Toast.LENGTH_SHORT).show();
19.                 imageView.setImageResource((imageAdapter.mImageIds)[position]);
20.             }
21.         });
22.     }
23. }
```

注意上述程序代码中第 9、11 行的 final 关键字必须要加上，否则程序会出错。第 17、18 行表示定义了一个 Toast，当用户单击了某个图片可显示用户选择了第几幅图；第 19 行表示将选中缩略图的放大图显示在下方。

运行 chap06_7 应用，界面如图 6-21 所示。

图 6-21　chap06_7 运行界面（一）

选中上图中任何一张图后单击，可看到如图 6-22 所示的界面。

【知识学习】GridView 组件

GridView 和 ListView 都是 AbsListView 的子类，因此二者在功能及用法上具有高度相似性，它们都是列表项。二者唯一的区别在于 ListView 只显示一列，而 GridView 可以显示多列。用户可以将 ListView 看成是 GridView 的特殊形式。

GridView 也需要通过 Adapter 进行数据绑定，可以调用 setAdapter 方法，用法和 ListView 基本一致。GridView 是以网格形式排列所包含的内容，每个单元格的内容可以是任意一个 View 组件；GridView 也可以使用 OnItemClickListener 及 OnItemSelectedListener 监听器事件。

GridView 常用的 XML 属性如下。

1）android：columnWidth：设置列的宽度，对应的方法为 setColumnWidth(int)。

图 6-22 chap06_7 运行界面（二）

2）android：gravity：设置对齐方式，对应的方法是 setGravity(int)。

3）android：horizontalSpacing：设置各元素之间的水平距离，对应的方法是 setHorizontalSpacing(int)；

4）android：numColumns：设置列数，对应的方法是 setNumColumns（int）。如果将该值改为 1，则 GridView 变成了 ListView。

5）android：stretchMode：设置拉伸模式，值有 NO_STRETCH，表示不拉伸；STRETCH_SPACING，表示拉伸元素之间的距离；STRETCH_SPACING_UNIFORM，将表格大小、表格元素之间的距离一起拉伸；STRETCH_COLUMN_WIDTH，仅拉伸表格大小。其对应的方法是 setStretchMode(int)。

任务 2　用 AdapterViewFlipper 实现自动播放图片

【任务描述】

用 AdapterViewFlipper 实现自动播放图片。

【任务实施】

1）创建一个 Android 应用，名称为"chap06_8"，项目名称为"AdapterViewFlipperTest"，Activity 名称为"AdapterViewFlipperActivity"文件，布局文件名称为"main"。

2）打开 main，其相应程序代码如下：

```
1.    < RelativeLayout xmlns:android = "http://schemas.android.com/apk/res/android"
2.        android:layout_width = "match_parent"
3.        android:layout_height = "match_parent" >
4.    < AdapterViewFlipper android:id = "@ + id/flipper"
5.        android:layout_width = "match_parent"
```

```
6.        android:layout_height = "match_parent"
7.        android:flipInterval = "5000"
8.        android:layout_alignParentTop = "true"/>
9.    <Button android:layout_width = "wrap_content"
10.       android:layout_height = "wrap_content"
11.       android:layout_alignParentBottom = "true"
12.       android:layout_alignParentLeft = "true"
13.       android:onClick = "prev"
14.       android:text = "上一个"/>
15.   <Button android:layout_width = "wrap_content"
16.       android:layout_height = "wrap_content"
17.       android:layout_alignParentBottom = "true"
18.       android:layout_centerHorizontal = "true"
19.       android:onClick = "next"
20.       android:text = "下一个"/>
21.   <Button android:layout_width = "wrap_content"
22.       android:layout_height = "wrap_content"
23.       android:layout_alignParentBottom = "true"
24.       android:layout_alignParentRight = "true"
25.       android:onClick = "auto"
26.       android:text = "自动播放"/>"
27. </RelativeLayout>
```

3) 打开 AdapterViewFlipperActivity, 其相应程序代码如下:

```
1.  public class AdapterViewFlipperActivity extends Activity {
2.      int[] imageIds = new int[] {
3.              R.drawable.img1,
4.              R.drawable.img2,
5.              R.drawable.img3,
6.              R.drawable.img4,
7.              R.drawable.img5,
8.              R.drawable.img6,
9.              R.drawable.img7,
10.             R.drawable.img8
11.     };
12.     AdapterViewFlipper flipper;
13.     @Override
14.     protected void onCreate(Bundle savedInstanceState) {
15.         super.onCreate(savedInstanceState);
16.         setContentView(R.layout.main);
17.         flipper = (AdapterViewFlipper)findViewById(R.id.flipper);
18.         //创建一个 BaseAdapter 对象,该对象负责提供 Gallery 所显示的列表项
```

```
19.         BaseAdapter adapter = new BaseAdapter()
20.         {
21.             public int getCount(){
22.                 return imageIds.length;
23.             }
24.             public Object getItem(int position){
25.                 return position;
26.             }
27.             public long getItemId(int position){
28.                 return position;
29.             }
30.             //该方法返回的 View 代表了每个列表项
31.             public View getView(int position,View convertView,ViewGroup parent){
32.                 //创建一个 ImageView
33.                 ImageView imageView = new ImageView(AdapterViewFlipperActivity.this);
34.                 imageView.setImageResource(imageIds[position]);
35.                 //设置 ImageView 的缩放类型
36.                 imageView.setScaleType(ImageView.ScaleType.FIT_XY);
37.                 //为 ImageView 设置布局参数
38.                 imageView.setLayoutParams(new LayoutParams(LayoutParams.MATCH_PARENT,
39.                     LayoutParams.MATCH_PARENT));
40.                 return imageView;
41.             }
42.         };
43.         flipper.setAdapter(adapter);
44.     }
45.
46.     public void prev(View source)
47.     {
48.         //显示上一个组件
49.         flipper.showPrevious();
50.         //停止自动播放
51.         flipper.stopFlipping();
52.     }
53.     public void next(View source)
54.     {
55.         //显示一个组件
56.         flipper.showNext();
57.         //停止自动播放
58.         flipper.stopFlipping();
59.     }
60.     public void auto(View source)
```

```
61.    }
62.        //开始自动播放
63.        flipper.startFlipping();
64.    }
65.    public boolean onCreateOptionsMenu(Menu menu){
66.        getMenuInflater().inflate(R.menu.adapter_view_flipper,menu);
67.        return true;
68.    }
69. }
```

运行项目 chap06_8，界面如图 6-23 所示。单击"上一个"按钮，可以查看上一张图片；单击"自动播放"按钮，可以自动播放图片。

【知识学习】AdapterViewFlipper 组件

AdapterViewFlipper 与 Gallery 都可以用来展示图片，但是 Gallery 是一次性把所有图片都预加载到屏幕上，而 AdapterViewFlipper 一次只加载一张图片。Gallery 在新版本的 Android 中不推荐使用。

AdapterViewFliper 继承自 AdapterViewAnimator，它可以显示 Adapter 提供的多个 View 组件，但每次只能显示一个 View 组件，开发者可通过 showPrevious 和 showNext 方法控制该组件显示上一个、下一个组件。

AdapterViewFlipper 可以在多个 View 切换过程中使用渐隐渐现的动画效果，还可以调用其 startFlipping 方法控制它"自动播放"下一个 View 组件。

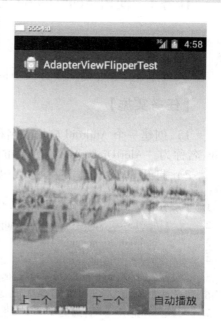

图 6-23 chap06_8 运行界面

AdapterViewFlipper 经常使用的属性如下。

1）android：animateFirstView：设置显示该组件的第一个 View 时是否使用动画。
2）android：inAnimation：设置组件显示时使用动画。
3）android：loopViews：设置是否循环。
4）android：outAnimation：设置组件隐藏时使用的动画。
5）android：autoStart：设置该组件是否自动播放，其对应的方法是 startFlipping。
6）android：flipInterval：设置自动播放的时间间隔，其对应的方法是 setFlipInterval(int)。

模块四 对话框/日期选择框等常用对话框的创建

【模块描述】

AlertDialog、DatePickerDialog 及 TabHost 组件在 App 应用中经常使用，本模块将基于这 3 个组件实现对话框、日期选择框及手机通话记录界面。

知识点	技能点
➢ AlertDialog 的创建及相关属性和方法 ➢ DatePickerDialog 的创建及相关属性和方法 ➢ TabHost 的相关属性和方法	➢ 使用 AlertDialog 组件实现对话框 ➢ 使用 DatePickerDialog 组件实现日期选择框 ➢ 通过 TabHost 选项卡模拟手机通话记录界面

任务1 使用 AlertDialog 实现对话框

【任务描述】

使用 AlertDialog 创建各种对话框。

【任务实施】

1）创建一个 Android 应用,名称为"chap06_9",项目名称为"AlertDialogTest",Activity 名称为"AlertDialogActivity",布局文件名称为"main"。

2）打开 main,其相应程序代码如下:

```
1.   <?xml version = "1.0" encoding = "utf-8"?>
2.   <LinearLayout xmlns:android = "http://schemas.android.com/apk/res/android"
3.       android:orientation = "vertical"
4.       android:layout_width = "match_parent"
5.       android:layout_height = "match_parent"
6.       android:gravity = "center_horizontal" >
7.   <!-- 显示一个普通的文本编辑框组件 -->
8.   <EditText
9.       android:id = "@+id/show"
10.      android:layout_width = "match_parent"
11.      android:layout_height = "wrap_content"
12.      android:editable = "false"/ >
13.  <!-- 定义一个普通的按钮组件 -->
14.  <Button
15.      android:layout_width = "match_parent"
16.      android:layout_height = "wrap_content"
17.      android:text = "简单对话框"
18.      android:onClick = "simple"/ >
19.  <!-- 定义一个普通的按钮组件 -->
20.  <Button
21.      android:layout_width = "match_parent"
22.      android:layout_height = "wrap_content"
23.      android:text = "简单列表项对话框"
```

```
24.         android:onClick = "simpleList"/ >
25.     <!--定义一个普通的按钮组件 -->
26.     < Button
27.         android:layout_width = "match_parent"
28.         android:layout_height = "wrap_content"
29.         android:text = "单选列表项对话框"
30.         android:onClick = "singleChoice"/ >
31.     <!--定义一个普通的按钮组件 -->
32.     < Button
33.         android:layout_width = "match_parent"
34.         android:layout_height = "wrap_content"
35.         android:text = "多选列表项对话框"
36.         android:onClick = "multiChoice"/ >
37.     <!--定义一个普通的按钮组件 -->
38.     < Button
39.         android:layout_width = "match_parent"
40.         android:layout_height = "wrap_content"
41.         android:text = "自定义列表项对话框"
42.         android:onClick = "customList"/ >
43.     <!--定义一个普通的按钮组件 -->
44.     < Button
45.         android:layout_width = "match_parent"
46.         android:layout_height = "wrap_content"
47.         android:text = "自定义 View 对话框"
48.         android:onClick = "customView"/ >
49. </LinearLayout >
50.     </LinearLayout >
51. </TabHost >
```

上述代码可以实现 6 种对话框,下面主要实现第一种对话框。

3) 打开 AlertDialogActivity,其相应程序代码如下:

```
1.  public class AlertDialogActivity extends Activity
2.  {
3.      TextView show;
4.      String[ ] items = new String[ ] {
5.              "Java 面向对象编程","数据库技术应用",
6.              "数据结构",
7.              "JSP 技术内幕" };
8.      public void onCreate( Bundle savedInstanceState)
9.      {
10.         super. onCreate( savedInstanceState) ;
```

```
11.        setContentView(R.layout.main);
12.        show = (TextView)findViewById(R.id.show);
13.    }
14.    public void simple(View source)
15.    {
16.        AlertDialog.Builder builder = new AlertDialog.Builder(this)
17.            //设置对话框标题
18.            .setTitle("简单对话框")
19.            //设置图标
20.            .setIcon(R.drawable.tools)
21.            .setMessage("对话框的测试内容\n第二行内容");
22.            //为 AlertDialog.Builder 添加"确定"按钮
23.            setPositiveButton(builder);
24.            //为 AlertDialog.Builder 添加"取消"按钮
25.            setNegativeButton(builder)
26.                .create()
27.                .show();
28.    }
29.    private AlertDialog.Builder setPositiveButton(
30.            AlertDialog.Builder builder)
31.    {
32.        //调用 setPositiveButton 方法添加"确定"按钮
33.        return builder.setPositiveButton("确定", new OnClickListener()
34.        {
35.            @Override
36.            public void onClick(DialogInterface dialog, int which)
37.            {
38.                show.setText("单击了"确定"按钮!");
39.            }
40.        });
41.    }
42.    private AlertDialog.Builder setNegativeButton(
43.            AlertDialog.Builder builder)
44.    {
45.        //调用 setNegativeButton 方法添加"取消"按钮
46.        return builder.setNegativeButton("取消", new OnClickListener()
47.        {
48.            @Override
49.            public void onClick(DialogInterface dialog, int which)
50.            {
51.                show.setText("单击了"取消"按钮!");
52.            }
```

```
53.              });
54.         }
55.    …
56. }
```

上述程序代码的第 14 ～ 28 行定义了按钮 simple 的单击事件方法，在这个方法中定义了 AlertDialog 对话框，并设置了相关内容；同时在代码的 29 ～ 41 行及 42 ～ 54 行定义了对话框中的"确定"及"取消"按钮的事件方法。

运行 chap06_9 应用，可得运行界面如图 6-24 所示。

单击上图中的"简单对话框"按钮，可见如图 6-25 所示的运行界面。

图 6-24　chap06_9 运行界面（一）

图 6-25　chap06_9 运行界面（二）

任务 2　使用 DatePickerDialog 实现日期输入

【任务描述】

在应用开发时，通常需要用户输入日期，而日期格式繁多，因此为了对用户日期输入的规范化，需要使用 DatePickerDialog 实现用户的日期输入。

【任务实施】

1）创建一个 Android 应用，名称为"chap06_10"，项目名称为"DatePickerDialogTest"，Activity 名称为"DatePickerDialogActivity"，布局文件名称为"main"。

2）打开 main，其相应程序代码如下：

```
1.   <?xml version = "1.0"  encoding = "utf-8"?>
2.   <LinearLayout xmlns:android = "http://schemas.android.com/apk/res/android"
```

```
3.          android:layout_width = "match_parent"
4.          android:layout_height = "match_parent"
5.          android:orientation = "vertical"
6.          android:background = "#ffffffff" >
7.      <TextView
8.          android:layout_width = "wrap_content"
9.          android:layout_height = "wrap_content"
10.         android:id = "@ + id/showtime"
11.         android:textColor = "#ff000000"
12.         android:text = " " />
13.     <Button
14.         android:layout_width = "wrap_content"
15.         android:layout_height = "wrap_content"
16.         android:id = "@ + id/setdate"
17.         android:text = "@ string/setdate" />
18. </LinearLayout>
```

3) 打开 DatePickerDialogActivity，其相应程序代码如下：

```
1.  /**
2.   *
3.   * DatePickerDialog 用以设置日期对话框,通过 OnDateSetListener 监听并重新设置日期,
4.   * 当日期被重置后,会执行 OnDateSetLintener 类中的方法 onDateSet
5.   *
6.   */
7.  public class DatePickerDialogActivity extends Activity {
8.      private TextView showdate;
9.      private Button setdate;
10.     private int year;
11.     private int month;
12.     private int day;
13.     public void onCreate( Bundle savedInstanceState)
14.     {
15.         super.onCreate(savedInstanceState);
16.         setContentView(R.layout.main);
17.         showdate = (TextView)this.findViewById(R.id.showtime);
18.         setdate = (Button)this.findViewById(R.id.setdate);
19.         //初始化 Calendar 日历对象
20.         Calendar mycalendar = Calendar.getInstance(Locale.CHINA);
21.         Date mydate = new Date();                    //获取当前日期 Date 对象
22.         mycalendar.setTime(mydate);                  //为 Calendar 对象设置时间为当前日期
23.         year = mycalendar.get(Calendar.YEAR);        //获取 Calendar 对象中的年
```

```
24.         month = mycalendar.get(Calendar.MONTH);        //获取Calendar对象中的月
25.         day = mycalendar.get(Calendar.DAY_OF_MONTH);    //获取这个月的第几天
26.         showdate.setText("当前日期:" + year + " - " + (month + 1) + " - " + day);//显示当前的年月日
27.         //添加单击事件——设置日期
28.         setdate.setOnClickListener(new OnClickListener(){
29.             @Override
30.             public void onClick(View v)
31.             {
32.                 /**
33.                  * 构造函数原型:
34.                  * public DatePickerDialog(Context context,
35.                  *         DatePickerDialog.OnDateSetListener callBack,
36.                  * int year,int monthOfYear,int dayOfMonth)
37.                  * 用content组件运行Activity,
38.                  * DatePickerDialog.OnDateSetListener:选择日期事件
39.                  * year:当前组件上显示的年;monthOfYear:当前组件上显示的月;
40.                  *     dayOfMonth:当前组件上显示的第几天
41.                  *
42.                  */
43.                 //创建DatePickerDialog对象
44.                 DatePickerDialog dpd = new
45. DatePickerDialog(DatePickerDialogActivity.this,Datelistener,year,month,day);
46.                 dpd.show();//显示DatePickerDialog组件
47.             }
48.         });
49.     }
50.     private DatePickerDialog.OnDateSetListener Datelistener = new
51. DatePickerDialog.OnDateSetListener()
52.     {
53.         /** params:view:该事件关联的组件
54.          * params:myyear:当前选择的年
55.          * params:monthOfYear:当前选择的月
56.          * params:dayOfMonth:当前选择的日
57.          */
58.         public void onDateSet(DatePicker view,int myyear,int monthOfYear,int dayOfMonth){
59. //修改year、month、day的变量值,以便以后单击按钮时,DatePickerDialog上显示上一次
60.         修改后的值
61.             year = myyear;
62.             month = monthOfYear;
63.             day = dayOfMonth;
64.             //更新日期
```

```
65.            updateDate();
66.        }
67.    //当DatePickerDialog关闭时更新日期显示
68.     private void updateDate()
69.        {
70.          //在TextView上显示日期
71.          showdate.setText("你选择的日期:" + year + " - " + (month + 1) + " - " + day);
72.         }
73.     };
74.  }
```

运行chap06_10，运行界面如图6-26所示。

当单击上图中"设置日期"按钮时，弹出如图6-27所示的日期选择框，选择日期后单击Done按钮即可。

图6-26　chap06_10运行界面（a）

图6-27　chap06_10运行界面（b）

【知识学习】DatePickerDialog组件

1）可以通过重载DatePickerDialog的setTitle来设置个性的标题。相应程序代码如下：

```
1.  Public void setTitle(CharSequence title){
2.      java.text.DateFormat dataFormat = (DateFormat.getDateFormat(getContext()));
3.      mCalendar = Calendar.getInstance();
4.      //可得到"12/31/1969(Wed)"形式的日期格式
5.      String strTitle = dataFormat.format(mCalendar.getTime()) +
6.              DateFormat.format("(E)",mCalendar.getTime()).toString();
7.      super.setTitle(dataFormat.format(strTitle));
8.  }
```

2）通过 onDateChanged 回调函数监听时间的改变，当时间改变时，以下函数会被调用。

```
1.  public void onDateChanged(DatePicker view,int year,int month,int day){
2.  Log.i("hubin","onDateChanged");
3.  }
```

3）通过 updateDate（）设置年、月、日。

语句中用 Public void updateDate（int year,int monthOfYear,int dayOfMonth）形式来设置 DatePickerDialog 的年、月、日。

任务 3　使用 TabHost 选项卡模拟手机通话记录界面

【任务描述】

使用选项卡 TabHost 模拟手机通话记录界面。

【任务实施】

1）创建一个 Android 应用，名称为"chap06_11"，项目名称为"TabHostTest"，Activity 名称为"TabHostActivity"，布局文件名称为"main"。

2）打开 main，其相应程序代码如下：

```
1.  <?xml version = "1.0" encoding = "utf-8"?>
2.  <TabHost
3.      xmlns:android = "http://schemas.android.com/apk/res/android"
4.      android:id = "@android:id/tabhost"
5.      android:layout_width = "match_parent"
6.      android:layout_height = "match_parent"
7.      android:layout_weight = "1" >
8.      <LinearLayout
9.          android:layout_width = "match_parent"
10.         android:layout_height = "match_parent"
11.         android:orientation = "vertical" >
12.         <TabWidget
13.             android:id = "@android:id/tabs"
14.             android:layout_width = "match_parent"
15.             android:layout_height = "wrap_content"/>
16.         <FrameLayout
17.             android:id = "@android:id/tabcontent"
18.             android:layout_width = "match_parent"
19.             android:layout_height = "match_parent" >
20.             <!--定义第一个标签页的内容-->
21.             <LinearLayout
22.                 android:id = "@+id/tab01"
23.                 android:orientation = "vertical"
```

```
24.            android:layout_width = "fill_parent"
25.            android:layout_height = "fill_parent" >
26.            <TextView
27.                android:layout_width = "wrap_content"
28.                android:layout_height = "wrap_content"
29.                android:text = "爸爸 - 2014/7/12"
30.                android:textSize = "11pt" />
31.            <TextView
32.                android:layout_width = "wrap_content"
33.                android:layout_height = "wrap_content"
34.                android:text = "姐姐 - 2014/7/18"
35.                android:textSize = "11pt" />
36.        </LinearLayout>
37.        <!-- 定义第二个标签页的内容 -->
38.        <LinearLayout
39.            android:id = "@+id/tab02"
40.            android:orientation = "vertical"
41.            android:layout_width = "fill_parent"
42.            android:layout_height = "fill_parent" >
43.            <TextView
44.                android:layout_width = "wrap_content"
45.                android:layout_height = "wrap_content"
46.                android:text = "大学同学 A - 2014/07/12"
47.                android:textSize = "11pt" />
48.            <TextView
49.                android:layout_width = "wrap_content"
50.                android:layout_height = "wrap_content"
51.                android:text = "舅舅 - 2014/08/20"
52.                android:textSize = "11pt" />
53.        </LinearLayout>
54.        <!-- 第三个选项卡(标签页内容)是通过调用另外一个 Activity 实现的 -->
55.        </FrameLayout>
56.    </LinearLayout>
57. </TabHost>
```

通过上面程序中的第 4、13、17 行代码可以看出，TabHost 组件内部需要两个组件，即 TabWidget 和 FrameLayout。其中，TabWidget 定义选项卡的标题条，FrameLayout 用于"层叠"组合多个选项页面。除此之外，上述布局文件中的 3 个组件的 id 并不是开发者自己定义的，而是引用了 Android 系统中已有的 id，开发者不可以随便定义，必须为下述内容。

① TabHost 的 id 为：tabhost。

② TabWidget 的 id 为：tabs。

③ FrameLayout 的 id 为：tabcontent。

从上述程序代码可以看出，页面定义了3个选项卡，其中，tab01和tab02这两个选项卡是使用LinearLayout组件（这两个组件嵌在FrameLayout中）定义的，每个LinearLayout中定义了两个TextView组件；而第3个选项卡没有在XML中定义，根据上述程序代码中的第54行提示，第3个选项卡是调用了另外一个Activity实现的，具体可参见下面的打开TabHostActivity类文件部分的代码。

3）打开TabHostActivity类文件，其相应程序代码如下：

```
1.   public class TabHostActivity extends TabActivity
2.   {
3.      @Override
4.      public void onCreate(Bundle savedInstanceState)
5.      {
6.          super.onCreate(savedInstanceState);
7.          setContentView(R.layout.main);
8.          //获取该Activity里面的TabHost组件
9.          TabHost tabHost = getTabHost();
10.
11.         //1. 创建第1个Tab页
12.         TabSpec tab1 = tabHost.newTabSpec("tab1")
13.             .setIndicator("已接电话")//设置标题
14.             .setContent(R.id.tab01);//设置内容
15.         //添加第1个标签页
16.         tabHost.addTab(tab1);
17.
18.         //2. 创建第2个标签
19.         TabSpec tab2 = tabHost.newTabSpec("tab2")
20.             //在标签标题上放置图标
21.             .setIndicator("呼出电话",getResources()
22.             .getDrawable(R.drawable.ic_launcher))
23.             .setContent(R.id.tab02);
24.         //添加第2个标签页
25.         tabHost.addTab(tab2);
26.
27.         //3. 创建第3个标签
28.         Intent intentTab3 = new Intent(TabHostActivity.this,NoneTelActivity.class);
29.         TabSpec tab3 = tabHost.newTabSpec("tab3")
30.             .setIndicator("未接电话")
31.             .setContent(intentTab3);
32.         //添加第3个标签页
33.         tabHost.addTab(tab3);
34.     }
35.  }
```

上述程序代码的第 11～16 行、18～25 行分别将布局文件中创建的两个 Tab 页面即 tab01 和 tab02 添加到 TabHost 容器中；第 27～33 行的程序代码表示创建了一个 Intent 对象 intentTab3，然后将这个 intentTab3 添加到第 3 个标签页中。

4）创建 Activity "NoneTelActivity.java" 及布局文件 nonetel.xml，其中，布局文件 XML 中的内容和 main.xml 中定义的标签页的内容相同，即 id 为 tab03。

运行 chap06_11 应用，运行界面如图 6-28 所示。

【练习】

1. 对项目 3 中的习题"模仿实现 QQ 登录布局界面"继续修改，要求如下：

1）当用户没有输入账号并单击"登录"按钮时，提示用户"请输入账号"的信息（可以采用 Toast 或弹出对话框等形式）；

2）当用户没有输入密码并单击"登录"按钮时，提示用户"请输入密码"信息（可以采用 Toast 或弹出对话框等形式）。

2. 创建一个图像浏览程序，界面整体布局如图 6-29 所示。

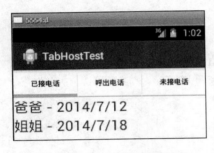

图 6-28　chap06_11 运行界面

图 6-29　图像浏览程序界面整体布局

说明：

1）相册列表以文件夹 1、文件夹 2 等字样来模拟相册数据的集合，并以 ListView 显示；

2）当前项目图像列表以 Gallery 的形式显示项目 Drawable 中的图像（读者可以对 Gallery 进行自学）；

3）图像区域以大图的形式显示上方 Gallery 的当前图像；

4）根据自己对图像浏览器应用的理解添加对话框等功能。

3. 请继续完成简单列表项对话框、单选列表项对话框、多选列表项对话框、自定义列表项对话框及自定义 View 对话框内容。

4. GridView 中各控件的联动实现：在 Activity 内添加一个 GridView，GridView 的子视图中包含一个 Button 和一个 TextView，在单击 Button 按钮后，会弹出一个 Toast 提示框并在之中显示 TextView 中的内容。预期效果如图 6-30

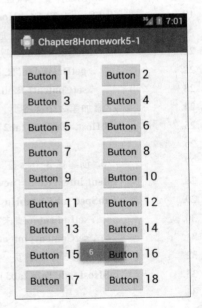

图 6-30　GridView 联动效果图

所示。当单击 TextView 为 6 的子视图中的 Button 按钮后弹出的提示框内容也为 6。

提示：在 getView 方法中，添加 Button 控件的监听事件，可考虑从父控件中获取子控件实例（此案例中以获取 TextView 控件实例）。

5. 仿微信、新浪微博标签栏的实现：在 Activity 界面中实现微信、新浪微博等应用中的底部标签栏，具体实现效果如图 6-31 所示。

图 6-31　仿微信、新浪微博标签栏的实现

提示：界面底部为 RadioGroup，将 RadioGroup 控件添加到 TabHost 中，更改 RadioGroup 中的 RadioButton 的 selector 属性配置，实现 RadioButton 在按下和非按下情况下的背景图片。使用 TabHost 中的 setCurrentTabByTag 方法可以在 RadioButton 的监听事件中切换 TabHost 中的 Tab 标签。

6. ListView 实现分页效果：在 ListView 中添加 20 条测试数据，要求 ListView 每次显示 5 条数据，并在单击"加载"按钮后，显示后 5 条数据，实现 ListView 的分页效果，如图 6-32 所示。

图 6-32　ListView 实现分页效果

单击"加载"按钮后实现分页，可显示后 5 条数据的信息。

提示：

1）在 BaseAdapter 中定义两个 List，一个存放所有的原始数据，另一个存放在 ListView 上的显示数据，并修改 getCount 返回值。

2）在单击"加载"按钮后，从原始数据中取出后 5 条数据存放于负责显示的 ListView 的 List 中，并通知 ListView 数据已更新。

项目七 Android 数据存储与 SQLite 数据库应用

本项目首先主要介绍了应用程序首选项 SharedPreferences 及如何读取首选项中的数据；接下来介绍了文件存储，详细介绍了 Android 内部存储数据及外部存储数据的读/写方法；最后重点讲述了 SQLite 数据库，详细介绍了使用 SQLiteDatabase 类和 SQLiteOpenHelper 类操作数据的步骤和方法，并讲解了 SQLite3 工具的用法。

【知识目标】

- SharedPreferences 的功能和作用
- Android 文件的存储
- 内部存储数据的读/写方法
- SD 卡数据读/写方法
- SQLiteDatabase 类的用法
- SQLiteOpenHelper 类的用法

【模块分解】

- 使用 SharedPreferences 存储简单数据
- 使用 FileInputStream 和 FileOutputStream 对 Android 内存数据进行读/写
- 使用 FileInputStream 和 FileOutputStream 对 SD 卡中的数据进行读/写
- 使用 SQLiteDatabase 类实现对 SQLite 数据库的 CRUD 操作
- 使用 SQLiteOpenHelper 类实现对 SQLite 数据库的 CRUD 操作
- 使用 SQLite3 工具操作数据库

模块一 Android 数据存储操作

【模块描述】

Android 应用中也需要数据存储，如果用户只保存少许的数据，如一些简单的字符串，那么可以使用普通的文件如 XML 文件进行；如果需要保存大量的数据需要使用 Android 提供的 SQLite 数据库；如果需要保存各种配置信息，如是否打开音效、是否使用震动效果等，可以保存在 Android 提供的首选项即 SharedPreferences 中。本模块将实现如下操作：

1）基于 SharedPreferences 实现系统参数设置；
2）基于 FileInputStream、FileOutputStream 实现内存数据读/写操作；
3）基于 FileInputStream、FileOutputStream 实现 SD 卡读/写操作。

知识点	技能点
➢ SharedPreferences 定义、相关属性和方法 ➢ SharedPreferences. Editor 相关方法 ➢ FileInputStream 输入流相关方法 ➢ FileOutputStream 输出流相关方法 ➢ 文件打开的模式	➢ 基于 SharedPreferences 实现数据的读/写操作 ➢ 基于 Environment 判断 SD 卡的操作 ➢ 内存及 SD 卡中数据的读/写操作 ➢ 创建 SD 卡镜像文件的步骤和方法

任务1 使用 SharedPreferences 设置系统参数

【任务描述】

如果只保存少许的数据,如一些简单的字符串,那么可以使用普通的文件如 XML 文件进行,此时可以使用 SharedPreferences 实现这个功能。

【任务实施】

1) 创建 Android 应用,名称为"chap07_1",项目名称为"SharedPreferencesTest",Activity 的名称为"SharedPreferencesActivity",布局文件名称为 main.xml。修改"main.xml"文件,增加两个 id 分别为 read 和 write 的按钮。

2) 打开 SharedPreferencesActivity,其相应程序代码如下:

```
1.   public class SharedPreferencesActivity extends Activity {
2.       SharedPreferences preferences;
3.       SharedPreferences. Editor editor;
4.       @Override
5.       protected void onCreate( Bundle savedInstanceState) {
6.           super. onCreate( savedInstanceState) ;
7.           setContentView( R. layout. main) ;
8.           preferences = this. getSharedPreferences( "shl" ,
9.                   Context. MODE_WORLD_READABLE) ;
10.          editor = preferences. edit( ) ;
11.      }
12.      public void readBtnClick( View v)
13.      {
14.          String time = preferences. getString( "time" ,null) ;
15.          int randNum = preferences. getInt( "random" ,0) ;
16.          String result = time == null?"SharedPreferences 中还没有被写入数据":
17.                  "写入时间:" + time + ",上次生成的随机数是:" + randNum;
18.          Toast. makeText( SharedPreferencesActivity. this ,result ,4000) . show( ) ;
19.      }
20.
21.      public void writeBtnClick( View v)
```

```
22.        {
23.            SimpleDateFormat date = new SimpleDateFormat("yyyy 年 MM 月 dd 日" + "hh:mm:ss");
24.            editor.putString("time",date.format(new Date()));
25.            editor.putInt("radom",(int)(Math.random()*100));
26.        editor.commit();
27.        }
28. }
```

运行应用 chap07_1，可得到如图 7-1 所示的界面。

单击上图中的"写入数据"按钮，然后单击"读取数据"按钮，运行界面如图 7-2 所示。

图 7-1　chap07_1 运行界面（a）

图 7-2　chap07_1 运行界面（b）

此时，程序完成了向 SharedPreferences 中写入数据的过程。想知道数据是存储在什么位置吗？请打开 DDMS 的 File Explorer 面板，展开文件浏览树，如图 7-3 所示。

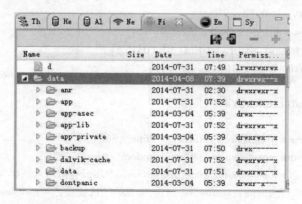
图 7-3　SharedPreferences 存储目录（一）

找到"data"根目录，继续打开根目录"data"下的文件夹"data"，找到以应用的包名命名的文件夹，如图 7-4 所示。

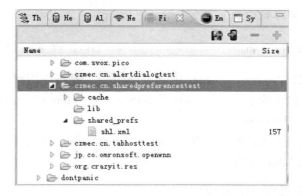

图 7-4 SharedPreferences 存储目录（二）

找到"shared_prefs"文件夹，可以看到一个 XML 文件"shl.xml"，这就是 SharedPreferences 的数据存储文件。选中这个文件，单击 按钮，如图 7-5 所示，可以将文件导出。

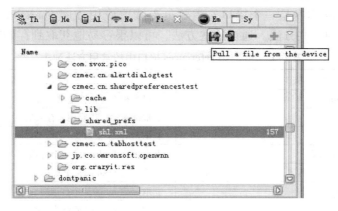

图 7-5 SharedPreferences 存储目录（三）

打开导出的 shl.xml 文件，其相应代码如下所示：

```
1.    <?xml version ='1.0'encoding ='utf-8'standalone ='yes'?>
2.    <map>
3.        <string name = "time">2014 年 07 月 31 日 07:57:05</string>
4.        <int name = "radom" value = "88" />
5.    </map>
```

可以发现，图 7-2 中显示的内容，就是从这个文件中读取的。

上面的案例是对应用程序本身的 SharedPreferences 进行读取，能否操作其他应用的 SharedPreferences 呢？

答案是肯定的。要操作其他应用的 SharedPreferences，需要首先指定该 SharedPreferences 的应用程序相应的访问权限，如上面的案例就是指定了 MODE_WORLD_READABLE 模式，表明该 SharedPreferences 数据可以被其他应用程序读取。

读取其他应用程序的 SharedPreferences 的一般步骤如下。

179

① 创建其他应用程序对应的上下文 Context，如代码：

createPackageContext("czmec.cn.sharedPreferencestest",Context.CONTEXT_IGNORE_SECURITY)

其中，第 1 个参数 "czmec.cn.sharedPreferencestest" 是指应用程序的程序包名。

② 调用其他应用程序的上下文 Context 的 getSharedPreferences(String name,int mode) 方法即可获取相应的 SharedPreferences 对象。

③ 如果需要向其他应用的 SharedPreferences 写入数据，只要调用 SharedPreferences 的 editor 方法获得相应的 Editor 对象即可。

【知识学习】应用程序首选项

Sharedpreferences 意思为 "共享首选项"。它是被所有 Activity 共享的，不能在应用程序包之外共享首选项。首选项以一组 "键/值" 的形式存储。而它所支持的数据类型有布尔值、浮点值、整型值、长整型值及字符串值等。

SharedPreferences 本身是一个接口，主要负责读取应用程序 SharedPreferences 数据，它包含如表 7-1 所示的一些方法。通过这些方法，可以很容易地获取到存储的内容。

表 7-1 SharedPreferences 接口中的常用方法列表

方 法	说 明
SharedPreferences.contains	通过名称查看该首选项是否存在
SharedPreferences.edit	获取编辑器，主要用来存入信息
SharedPreferences.getAll	获取所有键/值对
SharedPreferences.getBoolean	获取值为 Boolean 的某个首选项
SharedPreferences.getFloat	获取值为 Float 的某个首选项
SharedPreferences.getInt	获取值为 Int 的某个首选项
SharedPreferences.getLong	获取值为 Long 的某个首选项
SharedPreferences.getString	获取值为 String 的某个首选项

SharedPreferences 本身没有提供数据写入及写出的能力，实现数据的编辑功能主要通过 SharedPreferences 的内部接口。调用上表中的 SharedPreferences 的 edit 方法可以获取 Editor 编辑对象，通过对 Editor 的操作可以实现向 SharedPreferences 写入数据。Editor 对象中的方法及说明如表 7-2 所示。

表 7-2 SharedPreferences.Editor 的方法及说明

方 法	说 明
SharedPreferences.Editor.clear	移除所有首选项
SharedPreferences.Editor.remove	通过名称移除某个首选项
SharedPreferences.Editor.putBoolean	存入值为 Boolean 的首选项
SharedPreferences.Editor.putFloat	存入值为 Float 的首选项
SharedPreferences.Editor.putInt	存入值为 Int 的首选项
SharedPreferences.Editor.putLong	存入值为 Long 的首选项
SharedPreferences.Editor.putString	存入值为 String 的首选项
SharedPreferences.Editor.commit	提交此次编辑会话的所有更改

SharedPreferences 的使用步骤如下。

1）创建一个 SharedPreferences 对象的实例。

由于 SharedPreferences 本身是接口，因此无法创建 SharedPreferences 实例对象，一般通过下面的方法返回实例对象：

① preferences = getPreferences(MODE_PRIVATE)：获取 Activity 的私有首选项；

② sharedPres = getSharedPreferences("sharedPreferences", MODE_PRIVATE)：获取应用程序的首选项，其中 sharedPreferences 参数为实例名；第二个参数有以下 4 种情况。

a. Context.MODE_PRIVATE：默认模式，代表该文件为私有数据，只能被本应用程序读/写。此时如果写入文件，则会覆盖原有的文件内容。

b. Context.MODE_APPEND：追加模式，它会检查文件是否存在。如果存在，则在该文件中追加内容，而不是覆盖。

c. Context.MODE_WORLD_READABLE：表示当前文件可以被其他应用读取，但不能写。

d. Context.MODE_WORLD_WRITEABLE：表示当前文件可以被其他应用读/写。

2）创建一个 SharedPreferences.Editor 来修改首选项内容。

3）提交修改。

【任务拓展】

尝试读取上一个案例中所保存的 SharedPreferences 数据。

任务 2　Android 内存数据读/写操作

【任务描述】

基于 FileInputStream、FileOutputStream 实现内存数据读/写操作。

【任务实施】

1）创建 Android 应用，名称为"chap07_2"，项目名称为"InternalStorageTest"，Activity 名称为"InternalStorageActivity"，布局文件名称为 main.xml。修改"main.xml"文件，增加 3 个 id 分别为 readBtb、writeBtn、deleteBtn 的按钮，一个编辑框 edit1 和一个文本框 txt1。

2）打开 InternalStorageActivity，其相应程序代码如下：

```
1.    public class InternalStorageActivity extends Activity {
2.        EditText edit1;
3.        TextView edit2;
4.        final String FILE_NAME = "czmec.cn.internalstoragetest";
5.        protected void onCreate(Bundle savedInstanceState) {
6.            super.onCreate(savedInstanceState);
7.            setContentView(R.layout.main);
8.            edit1 = (EditText)this.findViewById(R.id.edit1);
9.            edit2 = (TextView)this.findViewById(R.id.txt1);
10.       }
11.       public void writeBtnClick(View v)
12.       {
```

```
13.         writeInternalStoragePrivate(FILE_NAME,edit1.getText().toString().getBytes());
14.         Toast.makeText(InternalStorageActivity.this,"写入成功",4000).show();
15.     }
16.     public void readBtnClick(View v)
17.     {
18.         String rtn = readInternalStoragePrivate(FILE_NAME);
19.         edit2.setText("读取的内容是:" + rtn.toString());
20.     }
21.     public void deleteBtnClick(View v)
22.     {
23.         deleteInternalStoragePrivate(FILE_NAME);
24.         Toast.makeText(InternalStorageActivity.this,"删除成功",4000).show();
25.     }
26.     …//相关功能方法
27. }
```

上述程序代码中的第 11 ~ 15 行定义了写入数据按钮的单击方法，调用了自行定义的方法 writeInternalStoragePrivate 实现数据的写入；第 16 ~ 20 行定义了读取数据按钮的单击方法，调用 readInternalStoragePrivate 方法实现数据的读取；第 21 ~ 25 行定义了删除按钮的单击方法，调用了 deleteInternalStoragePrivate 方法实现了数据的删除。这几个功能方法相应的程序代码如下。

```
1.  …
2.  /**
3.   *1. 将数据写入内部存储空间
4.   *@ param filename:文件名
5.   *@ param content:写入内容
6.   */
7.  public   void writeInternalStoragePrivate(
8.           String filename,byte[] content){
9.      try {
10.         FileOutputStream fos =
11.             openFileOutput(filename,Context.MODE_APPEND);
12.         fos.write(content);
13.         fos.close();
14.     } catch(FileNotFoundException e){
15.         e.printStackTrace();
16.     } catch(IOException e){
17.         e.printStackTrace();
18.     }
19. }
20. /**
21.  *2. 从内部存储空间文件中读/取数据
22.  *@ param filename:文件名称
23.  *@ return:文件内容
```

```
24.        */
25.      public String readInternalStoragePrivate(String filename){
26.          int len = 1024;
27.          byte[] buffer = new byte[len];
28.          try{
29.              FileInputStream fis = openFileInput(filename);
30.              int hasRead = 0;
31.              StringBuilder sb = new StringBuilder("");
32.              while((hasRead = fis.read(buffer)) > 0){
33.                  sb.append(new String(buffer,0,hasRead));
34.              }
35.              fis.close();
36.              return sb.toString();
37.          } catch(FileNotFoundException e){
38.              e.printStackTrace();
39.          } catch(IOException e){
40.              e.printStackTrace();
41.          }
42.          return null;
43.      }
44.      /**
45.       * 删除内部存储空间中的私有文件
46.       * @param filename:要删除的文件名称
47.       */
48.      public void deleteInternalStoragePrivate(String filename){
49.          File file = getFileStreamPath(filename);
50.          if(file != null){
51.              file.delete();
52.          }
```

运行 chap07_2 应用，可得到如图 7-6 所示的界面，在文本框中输入数据，然后单击"写入数据"按钮。单击"读取本地数据"按钮，可得到如图 7-7 所示的界面。

图 7-6　chap07_2 运行界面（一）

图 7-7　chap07_2 运行界面（二）

【知识学习】 文件存储

Java SE 开发者肯定使用过 Java SE 提供的功能强大的 I/O（输入/输出）流来实现文件存储，Android 应用开发者也可以使用强大的输入/输出流实现对手机存储器上文件的访问。

Android 文件存储介质分为内部存储（Internal Storage）和外部存储（External Storage）。手机内置的存储空间，称为内部存储，手机一旦出厂就无法改变，它也是手机的硬件指标之一。通常来讲，手机内置存储空间越大意味着手机价格会越贵（很多地方把它称为手机内存，这并不准确，内存是指手机运行时存储程序、数据和指令的地方；这里应该是对手机内部存储的简称，即内存，而并非严格意义上的内存）。

内部存储空间十分有限，因而显得可贵，所以要尽可能避免使用。另外，它也是系统本身和系统应用程序主要的数据存储所在地，一旦内部存储空间耗尽，手机也就无法使用了。上面所谈到的 SharedPreferences 和后面要谈到的 SQLite 数据库也都是存储在内部存储空间中的。而外部存储是指存储在外部的存储卡上，后面将进行详细的讲解。

Android 的 Activity 提供了 openFileOutput 方法，可以把数据输出到文件中。Activity 提供了如下一些方法。

1）FileInputStream openFileInput(String name)：打开应用程序数据文件夹下 name 文件对应的输入流。

2）FileOutputStream openFileOutput(String name, intmode)：打开应用程序数据文件夹下 name 文件对应的输出流。这个方法除了需要指定打开的 name 文件名外，还需要指定打开的文件模式，即 mode。该 mode 有如下几种值。

① Context. MODE_PRIVATE：为默认操作模式，代表该文件是私有数据，只能被应用本身访问，不能被其他应用程序读/写。在该模式下，写入的内容会覆盖原文件的内容。

② Context. MODE_APPEND：把新写入的内容追加到原文件中，此模式会检查文件是否存在，存在就往文件里追加内容，否则就创建新文件。

③ Context. MODE_WORLD_READABLE：表示当前文件可以被其他应用读取。

④ Context. MODE_WORLD_WRITEABLE：表示当前文件可以被其他应用读/写。Android 有一套自己的安全模型，应用程序（.apk）在安装时，系统就会分配给它一个 userid，当该应用要去访问其他资源比如文件的时候，就需要与 userid 匹配。默认情况下，任何应用创建的文件、SharedPreferences 或数据库都应该是私有的（位于/data/data//files 目录下），其他程序无法访问。除非在创建时指定了 MODE_WORLD_READABLE 或者 MODE_WORLD_WRITEABLE，只有这样其他程序才能正确访问。如果文件被其他应用读/写，可传入 openFileOutput("test. txt", Context. MODE_WORLD_READABLE + Context. MODE_WORLD_WRITEABLE)。

3）getDir(String name, int mode)：在应用程序数据文件夹下获取或创建 name 对应的子目录。

4）File getFilesDir：获取该应用程序数据文件夹的绝对路径。

5）String[] fileList：返回该应用程序数据文件夹下的全部文件。

6）deleteFile(String name)：删除该应用程序数据文件夹下的指定 name 文件。

任务3 Android SD 卡数据读/写操作

【任务描述】

使用 FileInputStream 和 FileOutputStream 对 SD 卡中的数据进行读/写。

【任务实施】

1）创建 Android 应用，名称为"chap07_3"，项目名称为"ExternalStorageTest"，Activity 名称为"ExternalStorageActivity"，布局文件名称为 main.xml。修改"main.xml"文件，增加两个 id 分别为 read、write 的按钮，两个 ID 为 edit1、edit2 的编辑框。

2）打开 ExternalStorageActivity，其相应程序代码如下。

```
1.   public class ExternalStorageActivity extends Activity {
2.       final String FILE = "/a";//文件名称,注意前面的"/"
3.       @Override
4.       public void onCreate(Bundle savedInstanceState) {
5.           super.onCreate(savedInstanceState);
6.           setContentView(R.layout.main);
7.
8.           Button read = (Button) findViewById(R.id.read);
9.           Button write = (Button) findViewById(R.id.write);
10.          //获取两个文本框
11.          final EditText edit1 = (EditText) findViewById(R.id.edit1);
12.          final EditText edit2 = (EditText) findViewById(R.id.edit2);
13.          //为 write 按钮绑定事件监听器
14.          write.setOnClickListener(new OnClickListener() {
15.              @Override
16.              public void onClick(View source) {
17.                  //将 edit1 中的内容写入文件中
18.                  writeToExternalStoragePublic(FILE,edit1.getText().toString().getBytes());
19.                  edit1.setText("");
20.                  Toast.makeText(ExternalStorageActivity.this,"写入成功",
21.  4000).show();
22.              }
23.          });
24.
25.          read.setOnClickListener(new OnClickListener() {
26.              @Override
27.              public void onClick(View v) {
28.                  //读取指定文件中的内容,并显示出来
29.                  String rtn = readExternallStoragePublic(FILE);
30.                  edit2.setText(rtn);
31.              }
```

```
32.            });
33.        }
34.    …//其他功能代码
35. }
```

以上程序中,第 18 行及 29 行的代码分别调用了将数据写入到 SD 卡及从 SD 卡中取出数据的方法,下面的程序代码为这两个方法的具体实现。

```
1.  …
2.  /**
3.   * 写到外存储器 SD 卡中
4.   * @ param filename – 文件名
5.   * @ param content – 内容
6.   */
7.  public void writeToExternalStoragePublic(String filename,byte[ ] content){
8.      //String packageName = this.getPackageName();
9.      //String path = "/Android/data/" + packageName + "/files/";
10.     if(isExternalStorageAvailable() &&
11.         !isExternalStorageReadOnly()){
12.         try{
13.             //获取 SD 卡对应的存储目录
14.             File sdCardfileDir = Environment.getExternalStorageDirectory();
15.             File targetFile = new File(sdCardfileDir.getCanonicalPath() + filename);
16.             //以指定文件创建 RandomAddressFile 对象
17.             RandomAccessFile raf = new RandomAccessFile(targetFile,"rw");
18.             //将记录指针移动到文件末尾
19.             raf.seek(targetFile.length());
20.             //输出文件内容
21.             raf.write(content);
22.             //关闭 randomAccessFile
23.             raf.close();
24.
25.         }catch(FileNotFoundException e){
26.             e.printStackTrace();
27.         }catch(IOException e){
28.             e.printStackTrace();
29.         }
30.     }
31. }
32. /**
33.  * 从外部内存读取数据
34.  * @ param filename – 文件名
```

```
35.         * @return the file contents
36.         */
37.        public String readExternallStoragePublic(String filename){
38.            int len = 1024;
39.            byte[] buffer = new byte[len];
40.            //String packageName = this.getPackageName();
41.            //String path = "/Android/data/" + packageName + "/files/";
42.            if(!isExternalStorageReadOnly()){
43.                try{
44.                    //获取SD卡对应的存储目录
45.                    File sdCardfileDir = Environment.getExternalStorageDirectory();
46.                    FileInputStream fis = new
47.                        FileInputStream(sdCardfileDir.getCanonicalPath() + filename);
48.                    //将指定输入流包装成BufferedReader
49.                    BufferedReader br = new BufferedReader(new InputStreamReader(fis));
50.                    StringBuilder sb = new StringBuilder("");
51.                    String line = null;
52.                    //循环读取文件内容
53.                    while((line = br.readLine()) != null){
54.                        sb.append(line);
55.                    }
56.                    fis.close();
57.                    br.close();
58.                    return sb.toString();
59.                } catch(FileNotFoundException e){
60.                    e.printStackTrace();
61.                } catch(IOException e){
62.                    e.printStackTrace();
63.                }
64.            }
65.            return null;
66.        }
67.
68.        /**
69.         * 判断外存储器SD卡是否可用
70.         */
71.        public boolean isExternalStorageAvailable(){
72.            boolean state = false;
73.            String extStorageState = Environment.getExternalStorageState();
74.            if(Environment.MEDIA_MOUNTED.equals(extStorageState)){
75.                state = true;
76.            }
```

```
77.            return state;
78.        }
79.
80.        /**
81.         * 判断外部存储器是否只可读
82.         */
83.        public boolean isExternalStorageReadOnly() {
84.            boolean state = false;
85.            String extStorageState = Environment.getExternalStorageState();
86.            if (Environment.MEDIA_MOUNTED_READ_ONLY.equals(extStorageState)) {
87.                state = true;
88.            }
89.            return state;
90.        }
91.
92.        /**
93.         * 删除外部内存中的文件
94.         * @param filename - 文件名
95.         */
96.        public void deleteExternalStoragePublicFile(String filename) {
97.            //String packageName = this.getPackageName();
98.            //String path = "/Android/data/" + packageName + "/files/" + filename;
99.            //获取 SD 卡对应的存储目录
100.           File sdCardfileDir = Environment.getExternalStorageDirectory();
101.           File file;
102.           try {
103.               file = new File(sdCardfileDir.getCanonicalPath(), filename);
104.               if (file != null) {
105.                   file.delete();
106.               }
107.           } catch (IOException e) {
108.               //TODO Auto-generated catch block
109.               e.printStackTrace();
110.           }
111.       ...
```

上述代码中，第 73 行中的 getExternalStorageState 方法是用来获取外部存储的可用状态，可能的状态包括挂载成功、只读、找不到存储卡等，对应的状态值分别如下。

① MEDIA_UNMOUNTED：代表存储卡存在，但没有被挂载。

② MEDIA_UNMUNTABLE：代表存储卡存在，但无法被挂载。

③ MEDIA_SHARED：代表存储卡存在，但未被挂载。

④ MEDIA_REMOVED：代表存储卡不存在。

⑤ MEDIA_NOFS：代表存储卡存在，但是空白的，或者存储卡使用的文件系统格式不被手机系统支持。

⑥ MEDIA_MOUNTED_READ_ONLY：代表存储卡以只读模式挂载了。

⑦ MEDIA_MOUNTED：代表存储卡被成功挂载。

⑧ MEDIA_CHECKING：代表存储卡正在被检查的过程中。

⑨ MEDIA_BAD_REMOVAL：代表存储卡在卸载前被拔掉了。

3）打开 AndroidManifest.xml 文件，添加读/写 SD 卡的权限，其程序如下所示：

```
1.    < uses – permission
2.       android:name = " android. permission. MOUNT_UNMOUNT_FILESYSTEMS"/>
3.    < uses – permission
4.       android:name = " android. permission. WRITE_EXTERNAL_STORAGE"/>
```

运行 chap07_3 应用，可得如图 7-8 所示的运行界面。

当单击上图中的"读取"按钮时，可以将存放在 SD 卡中的 a.txt 文件中的数据取出并显示在页面中。这个文件是放在 SD 卡的"/mnt/sdcard/"路径下的，打开 DDMS，找到如图 7-9 所示的路径。

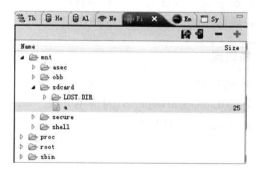

图 7-8　chap07_3 运行界面　　　　　图 7-9　SD 卡下的文件路径

选中图 7-9 中的 a 文件，再单击 按钮，可以将文件导出。

但是，读者可能已经发现，打开 DDMS 后，在/mnt/sdcard 目录下没有这个文件，这是什么原因呢？

需要指出的是，当开发者直接在 Eclipse 环境中通过模拟器运行 Android 应用时，模拟器中并没有 SD 卡。为了让 Eclipse 启动的 Android 模拟器有 SD 卡，可以通过如下方法实现。

① 创建一个 SD 卡镜像文件。

打开 cmd，找到 Android 安装的 SDK tools 路径，输入如图 7-10 所示的命令。

该命令会在当前目录下生成一个 sdcard.img 文件，该文件就是 Android 模拟器的 SD 卡镜像文件。其中，200 MB 为 SD 卡的容量，目前 Android 支持 8 MB ～ 128 GB 的 SD 卡。

图 7-10 输入创建 SD 卡镜像文件的命令

② 将生成的 sdcard.img 镜像文件复制到 E:\sdcard 文件夹下。在 Eclipse 环境下打开 Android 虚拟设备管理器，选中当前要使用的 AVD 模拟器，单击 Edit 按钮，进入 Android 虚拟设备配置窗口，在这个窗口中指定 SD 卡的镜像文件，如图 7-11 所示。

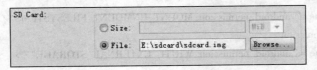

图 7-11 创建 SD 卡镜像文件

之后关闭模拟器，重新运行模拟器，这个模拟器中就带 SD 卡了。

模块二　SQLite 数据库操作

【模块描述】

Android 系统提供了一个小型的 SQLite 数据库，适用于在资源有限的设备上的适量数据存取，如手机、PDA 等。本模块将详细讲解基于 SQLiteDatabase 和 SQLiteOpenHelper 实现对 SQLite 数据库的增、删改、查操作。

知识点	技能点
➢ SQLiteDatabaseH 类的常用方法 ➢ SQLiteOpenHelper 类的常用方法 ➢ SQLite3 工具相关命令	➢ 使用 SQLiteDatabase 类实现 SQLite 数据库操作 ➢ 使用 SQLiteOpenHelper 实现 SQLite 数据库操作 ➢ 使用 SQLite3 工具操作数据库

任务 1　使用 SQLiteDatabase 实现 SQLite 数据库操作

【任务描述】

使用 SQLiteDatabase 类实现 SQLite 数据库操作。

【任务实施】

1）创建 Android 应用，名称为"chap07_4"，项目名称为"SQLiteDatabaseTest"，Activity 名称为"SQLiteDatabaseActivity"，布局文件名称为 main.xml。

2) 打开 "main.xml" 文件,相应程序内容如下:

```xml
1.   <LinearLayout xmlns:android="http://schemas.android.com/apk/res/android"
2.       android:layout_width="fill_parent"
3.       android:layout_height="fill_parent"
4.       android:layout_gravity="center_vertical"
5.       android:orientation="vertical">
6.       <LinearLayout
7.           android:layout_width="match_parent"
8.           android:layout_height="wrap_content"
9.           android:orientation="horizontal">
10.          <TextView
11.              android:layout_width="wrap_content"
12.              android:layout_height="wrap_content"
13.              android:text="用户名"/>
14.          <EditText
15.              android:id="@+id/userName"
16.              android:layout_width="match_parent"
17.              android:layout_height="wrap_content"/>
18.      </LinearLayout>
19.      <LinearLayout
20.          android:layout_width="match_parent"
21.          android:layout_height="wrap_content"
22.          android:orientation="horizontal">
23.          <TextView
24.              android:layout_width="wrap_content"
25.              android:layout_height="wrap_content"
26.              android:text="密 码"/>
27.          <EditText
28.              android:id="@+id/userPassword"
29.              android:layout_width="match_parent"
30.              android:layout_height="wrap_content"
31.              android:inputType="textPassword"/>
32.      </LinearLayout>
33.      <LinearLayout
34.          android:layout_width="match_parent"
35.          android:layout_height="wrap_content"
36.          android:layout_gravity="center_vertical"
37.          android:gravity="center"
38.          android:orientation="horizontal">
39.          <Button
40.              android:id="@+id/addBtn"
41.              android:layout_width="wrap_content"
```

```
42.            android:layout_height = "wrap_content"
43.            android:layout_gravity = "center_vertical"
44.            android:text = "插入数据"
45.            android:onClick = "addBtn"/>
46.        </LinearLayout>
47.        <LinearLayout
48.            android:layout_width = "match_parent"
49.            android:layout_height = "wrap_content"
50.            android:orientation = "vertical" >
51.            <ListView
52.                android:id = "@+id/display"
53.                android:layout_width = "match_parent"
54.                android:layout_height = "wrap_content" >
55.            </ListView>
56.        </LinearLayout>
57.    </LinearLayout>
```

3) 打开SQLiteDatabaseActivity,相应程序代码如下:

```
1.    public class SQLiteDatabaseActivity extends Activity
2.    {
3.        SQLiteDatabase db;
4.        Button bn = null;
5.        ListView listView;
6.        @Override
7.        public void onCreate(Bundle savedInstanceState)
8.        {
9.            super.onCreate(savedInstanceState);
10.           setContentView(R.layout.main);
11.           //创建或打开数据库(此处需要使用绝对路径)
12.           db = SQLiteDatabase.openOrCreateDatabase(this.getFilesDir().toString() +
13.               "/test1.db3",null);
14.           listView = (ListView)findViewById(R.id.display);
15.       }
16.       public void addBtn(View v)
17.       {
18.           //获取用户输入
19.           String userName = ((EditText)findViewById(R.id.userName)).getText().toString();
20.           String UserPassword = ((EditText)findViewById(R.id.userPassword)).getText().toString();
21.           String sqlSelect = "select * from UserInfo";
22.           try
23.           {
24.               insert(db,userName,UserPassword);
```

```
25.              Cursor cursor = db. rawQuery(sqlSelect,null);
26.              inflateList(cursor);
27.          }
28.          catch(SQLiteException se)
29.          {
30.              //执行 DDL 创建数据表
31.              String sqlCreate = "create table UserInfo(_id integer"
32.                      + "primary key autoincrement,"
33.                      + "userName varchar(20),"
34.                      + "userPassword varchar(50))";
35.              db. execSQL(sqlCreate);
36.              //执行 Insert 语句插入数据
37.              insert(db,userName,UserPassword);
38.              //执行查询
39.              Cursor cursor = db. rawQuery(sqlSelect,null);
40.              inflateList(cursor);
41.          }
42.      }
43.      private void insert(SQLiteDatabase db,String name,String pass)
44.      {
45.          //执行插入语句
46.          db. execSQL("insert into UserInfo values(null ,?,?)"
47.                  ,new String[]{name,pass});
48.      }
49.
50.      private void inflateList(Cursor cursor)
51.      {
52.          //填充 SimpleCursorAdapter
53.          SimpleCursorAdapter adapter = new SimpleCursorAdapter(
54.                  SQLiteDatabaseActivity. this,
55.                  R. layout. listitems,cursor,
56.                  new String[]{"userName","userPassword"}
57.                  ,new int[]{R. id. userName,R. id. userPassword},
58.                  CursorAdapter. FLAG_REGISTER_CONTENT_OBSERVER);
59.          //显示数据
60.          listView. setAdapter(adapter);
61.      }
62.
63.      @Override
64.      public void onDestroy()
65.      {
66.          super. onDestroy();
67.          //退出程序时关闭 SQLiteDatabase
68.          if(db != null && db. isOpen())
```

```
69.            }
70.               db.close();
71.            }
72.        }
73.   }
```

上述程序代码中的第 12、13 行创建或打开了 SQLite 数据库。当用户单击"插入数据"按钮后，代码第 43～48 行将会向表 UserInfo 中插入一行数据。同时，第 39 行代码将会执行查询语句，将数据库中所有的数据查询出来并返回一个 Cursor 对象，再使用第 40 行的 inflateList 方法将数据填充到 SimpleCursorAdapter 适配器中，并绑定 ListView 组件，将查询结果显示出来。请读者注意第 50～61 行中 inflateList 方法的实现。在定义 SimpleCursorAdapter 适配器时，其构造方法中的参数和前面讲解的 SimpleAdapter 类似，不同的是，前者的返回为一个 Cursor 对象。第 55 行的 R.layout.listitems 为第 2 个参数，是列表的布局，其中定义了两个 EditView 组件，其 id 分别为 userName 和 userPassword，这和 55 行需要绑定的列表项一致。

运行 chap07_4 应用，可得如图 7-12 所示的界面。

需要注意的是，通过上述代码的第 31 行可以发现，使用 SimpleCursorAdapter 封装 Cursor 时，要求数据库表

图 7-12　chap07_4 运行界面

的主键的列名必须为"_id"，因为 SimpleCursorAdapter 只能识别列名为_id 的主键。同样，数据库操作完毕后必须关掉，否则可能引发一些异常情况，这里第 64～72 行的代码通过重写 Activity 的 onDestroy 方法来实现的，当程序退出此 Activity 时就会回调该方法。

可见，使用 SQLiteDatabase 类对数据库进行操作的步骤一般如下。

① 获取 SQLiteDatabase 对象。
② 调用 SQLiteDatabase 中的方法来执行相关 SQL 语句。
③ 操作 SQL 语句的执行结果，如用 SimpleCursorAdapter 封装 Cursor 对象。
④ 关闭 SQLiteDatabase，回收相关资源。

【知识学习】**SQLiteDatabase 类**

Android 提供了一个名为 SQLiteDatabase 的类，开发者通过获取这个类的对象后就可以实现对 SQLite 数据库的操作。SQLiteDatabase 类中常用的数据库操作方法如下。

1) static SQLiteDatabase openOrCreateDatabase(String pah, SQLiteDatabase.CursorFactory factory)：打开或创建（如果不存在）path 文件所代表的 SQLite 数据库，CursorFactory 代表着在创建 Cursor 对象时使用的工厂类，如果为 null 的话，则使用默认的工厂类。

2) static SQLiteDatabase openOrCreateDatabase(File file, SQLiteDatabase.CursorFactory factory)：打开或创建（如果不存在）file 文件所代表的 SQLite 数据库。

3) static SQLiteDatabase openDatabase(String path, SQLiteDatabase.CursorFactory factory, int

flags)：打开 path 文件所代表的 SQLite 数据库。

上述 3 个方法是创建数据库或打开数据库的方法，由于是静态方法，因此可以直接通过 SQLiteDatabase 的类名访问。

当成功返回 SQLiteDatabase 数据库对象后，可以对数据库进行 CRUD 操作[①]，类 SQLiteDatabase 提供的方法很多，下面介绍几种常用的方法。

1) execSQL(String sql, Object[] bindArgs)：执行 SQL 语句，其中 SQL 语句可以带"?"占位符。此 SQL 语句包括 CRUD 等所有数据库操作的 SQL 语句，只能执行 DDL 语句或 DML 语句，如创建表结构、插入数据、删除数据、更新数据。

2) execSQL(String sql)：执行 SQL 语句。此 SQL 语句包括 CRUD 等所有数据库操作的 SQL 语句。

3) rawQuery(String sql, String[] selectionArgs)：执行带占位符的 SQL 查询语句。注意，只能执行查询语句。其返回结果是一个 Cursor 对象，类似于 JDBC 中的 ResultSet 结果集。对 Cursor 对象操作，可以通过 Cursor 中提供的方法实现。

4) beginTransaction：开始一个事务。

5) endTransaction：结束一个事务。

6) close：关闭数据库。当对数据库操作完成后需要关闭数据库，否则会抛出 SQLiteException 异常。

7) deleteDatabase(String path)：删除指定的数据库。

可以看到，SQLite 数据库也支持事务管理，本书不进行详细介绍，感兴趣的读者可以查阅其他相关参考书。

上述一些方法是常用的对数据库操作的方法，读者要使用上述方法实现对 SQLite 数据库的操作，必须要对 SQL 语法熟悉。考虑到部分开发者可能对 SQL 语法不熟悉等，Android 又提供了如下功能强大的增、删、改、查方法。

1) insert(String table, String nullColumnHack, ContentValues values)：向表中插入数据。其中，第 1 个参数为表的名字；第 2 个参数一般为 null；第 3 个参数是 ContentValues：这是一个键/值对，使用的时候需要 new 一个 ContentValues，然后使用 put 方法添加数据。插入若成功，返回新插入 row 的 id，失败则返回 -1。

2) update(String talbe, ContentValues vaules, String whereClause, String[] whereArgs)：更新指定表中的特定数据。

3) delete(String table, String whereClause, String[] whereArgs)：删除指定表中的特定数据。第 2 个参数为删除的条件，如果为 null 则全部删除；第 3 个参数为字符串数组，和 whereClause 配合使用。比如，如果 whereClause 的条件已经给出"name = " + num，num 是传入的参数，则 whereArgs 可设置为 null。又如，当 whereClause 中包含"?"时，则 whereArgs 这个数组中的值将依次替换 whereClause 中出现的"?"。

4) Cursor query(String table, String[] columns, String whereClause, String[] whereArgs, String groupBy, String having, String orderBy)：对数据表执行查询。

[①] CRUD 是在做计算处理时的增加（Create）、读取（Retrieve）、更新（Update）和删除（Delete）几个单词的首字母简写。主要被用描述软件系统中数据库或者持久层的基本操作功能。

rawQuery 方法和 query 方法返回的都是 Cursor 对象，开发者可以使用如下方法对 Cursor 对象进行操作，进而读取其中的数据。

1) move(int offset)：将记录指针向上或向下移动指定的行数。offset 参数为正数即向下移动，负数即向上移动。

2) boolean moveToFirst：将记录指针移动到第一行，移动成功返回 true。

3) boolean moveToLast：将记录指针移动最后一行，移动成功返回 true。

4) boolean moveToNext：将记录指针移动到下一行，移动成功返回 true。

5) boolean moveToPosition(int postion)：将记录指针移动到 postion 指定位置，移动成功返回 true。

6) boolean moveToPrevious：将记录指针移动到上一行，移动成功返回 true。

7) getXXX：一旦将记录指针移动到指定的行，开发者就可以调用 Cursor 的 getXXX() 方法获取该行指定的列的数据。

任务 2　使用 SQLiteOpenHelper 实现 SQLite 数据库操作

【任务描述】

使用 SQLiteOpenHelper 类实现 SQLite 数据库操作。

【任务实施】

1) 创建安卓应用，名称为 "chap07_5"，项目名称为 "SQLiteOpenHelperTest"，Activity 名称为 "SQLiteOpenHelperActivity"，布局文件名称为 "main.xml"。

2) 打开 "main.xml" 文件，其相应程序代码内容如下：

```
1.  <?xml version = "1.0" encoding = "utf - 8"?>
2.  <LinearLayout xmlns:android = "http://schemas.android.com/apk/res/android"
3.      android:layout_width = "fill_parent"
4.      android:layout_height = "fill_parent"
5.      android:orientation = "vertical" >
6.      <Button
7.          android:layout_width = "fill_parent"
8.          android:layout_height = "wrap_content"
9.          android:onClick = "add"
10.         android:text = "添加"/>
11.     <Button
12.         android:layout_width = "fill_parent"
13.         android:layout_height = "wrap_content"
14.         android:onClick = "update"
15.         android:text = "更新"/>
16.     <Button
17.         android:layout_width = "fill_parent"
18.         android:layout_height = "wrap_content"
19.         android:onClick = "delete"
```

```
20.         android:text = "删除"/>
21.     <Button
22.         android:layout_width = "fill_parent"
23.         android:layout_height = "wrap_content"
24.         android:onClick = "query"
25.         android:text = "查询"/>
26.     <Button
27.         android:layout_width = "fill_parent"
28.         android:layout_height = "wrap_content"
29.         android:onClick = "queryTheCursor"
30.         android:text = "查询Cursor"/>
31.     <ListView
32.         android:id = "@+id/listView"
33.         android:layout_width = "fill_parent"
34.         android:layout_height = "wrap_content"/>
35. </LinearLayout>
```

3) 创建DatabaseHelper类,继承SQLiteOpenHelper类,其相应程序代码如下:

```
1.  public class DatabaseHelper extends SQLiteOpenHelper//继承SQLiteOpenHelper类
2.  {
3.      //数据库版本号
4.      private static final int DATABASE_VERSION = 1;
5.      //数据库名
6.      private static final String DATABASE_NAME = "TestDB.db";
7.      //数据表名,一个数据库中可以有多个表(虽然本例中只建立了一个表)
8.      public static final String TABLE_NAME = "PersonTable";
9.      //构造函数,调用父类SQLiteOpenHelper的构造函数
10.     public DatabaseHelper(Context context,String name,CursorFactory factory,
11.             int version,DatabaseErrorHandler errorHandler)
12.     {
13.         super(context,name,factory,version,errorHandler);
14.     }
15.     public DatabaseHelper(Context context,String name,CursorFactory factory,
16.             int version)
17.     {
18.         super(context,name,factory,version);
19.         //SQLiteOpenHelper的构造函数参数
20.         //context:上下文环境
21.         //name:数据库名字
22.         //factory:游标工厂(可选)
23.         //version:数据库模型版本号
24.     }
25.
26.     public DatabaseHelper(Context context)
```

```
27.         }
28.             super(context,DATABASE_NAME,null,DATABASE_VERSION);
29.         //数据库实际是在 getWritableDatabase 或 getReadableDatabase 方法调用时被创建
30.             Log.d(AppConstants.LOG_TAG,"DatabaseHelper Constructor");
31.             //CursorFactory 设置为 null,使用系统默认的游标工厂
32.         }
33.         //继承 SQLiteOpenHelper 类,要覆写的 3 个方法:onCreate、onUpgrade、onOpen
34.         @Override
35.         public void onCreate(SQLiteDatabase db)
36.         {
37.             //调用时间:数据库第一次创建时,onCreate 方法会被调用
38.             //onCreate 方法有一个 SQLiteDatabase 对象作为参数,根据需要对这个对象填充表和初始化数据
39.             //这个方法主要用以完成创建数据库后对数据库的操作
40.             Log.d(AppConstants.LOG_TAG,"DatabaseHelper onCreate");
41.             //构建创建表的 SQL 语句(可以从 SQLite Expert 工具的 DDL 粘贴过来,加进 StringBuffer 中)
42.             StringBuffer sBuffer = new StringBuffer();
43.             sBuffer.append("CREATE TABLE [" + TABLE_NAME + "] (");
44.             sBuffer.append("[_id] INTEGER NOT NULL PRIMARY KEY AUTOINCREMENT,");
45.             sBuffer.append("[name] TEXT,");
46.             sBuffer.append("[age] INTEGER,");
47.             sBuffer.append("[info] TEXT)");
48.             //执行创建表的 SQL 语句
49.             db.execSQL(sBuffer.toString());
50.         //即使程序修改重新运行,只要数据库已经创建过,就不会再进入这个 onCreate 方法
51.         }
52.         public void onUpgrade(SQLiteDatabase db,int oldVersion,int newVersion)
53.         {
54.             //如果 DATABASE_VERSION 值被改为别的数,系统发现现有数据库版本不同,即会调用 onUpgrade
55.             //onUpgrade 方法的 3 个参数:一个 SQLiteDatabase 对象、一个旧的版本号和一个新的版本号
56.             //这样就可以把一个数据库从旧的模型转变到新的模型
57.             //这个方法主要完成更改数据库版本的操作
58.             Log.d(AppConstants.LOG_TAG,"DatabaseHelper onUpgrade");
59.             db.execSQL("DROP TABLE IF EXISTS" + TABLE_NAME);
60.             onCreate(db);
61.             //上述做法简单来说就是,通过检查常量值来决定如何在升级时删除旧表,然后调用 onCreate
62.         //来创建新表
63.             //一般在实际项目中是不能这么做的,正确的做法是在更新数据表结构时还要考虑用户存
64.         //放于数据库中的数据不丢失
65.         }
66.         public void onOpen(SQLiteDatabase db)
67.         {
68.             super.onOpen(db);
```

```
69.          //每次打开数据库之后首先被执行
70.          Log. d( AppConstants. LOG_TAG," DatabaseHelper onOpen") ;
71.      }
72. }
```

4) 定义一个 Person 类作为测试数据,其相应程序代码如下:

```
1.  public class Person
2.  {
3.      public int _id;
4.      public String name;
5.      public int age;
6.      public String info;
7.      public Person( )
8.      {
9.      }
10.     public Person( String name, int age, String info)
11.     {
12.         this. name = name;
13.         this. age = age;
14.         this. info = info;
15.     }
16. }
```

5) 定义一个管理类类,实现对各种操作的封装,其相应程序代码如下。

```
1.  public class DBManager
2.  {
3.      private DatabaseHelper helper;
4.      private SQLiteDatabase db;
5.      public DBManager( Context context)
6.      {
7.          Log. d( AppConstants. LOG_TAG," DBManager --> Constructor") ;
8.          helper = new DatabaseHelper( context) ;
9.          //因为 getWritableDatabase 内部调用了 mContext. openOrCreateDatabase( mName,0,
10.         //mFactory) ;
11.         //所以要确保 context 已初始化,这里可以把实例化 DBManager 的步骤放在 Activity 的
12. //onCreate 里
13.         db = helper. getWritableDatabase( ) ;
14.     }
15.     public void add( List < Person > persons)
16.     {
```

```
17.         Log. d( AppConstants. LOG_TAG ," DBManager --> add" );
18.         //采用事务处理,确保数据完整性
19.         db. beginTransaction( );//开始事务
20.         try
21.         {
22.             for ( Person person : persons)
23.             {
24.                 db. execSQL( " INSERT INTO" + DatabaseHelper. TABLE_NAME
25.                     + " VALUES( null,?,?,?)" ,new Object[ ]{person. name,
26.                     person. age , person. info} );
27.     //上述 execSQL 方法使用了"?"作为参数占位符,在使用时要注意"?"和参数值的顺序对应关系,
28.     //如第一个"?"对应数组中的 person. name
29.             }
30.             db. setTransactionSuccessful( );    //设置事务成功完成
31.         }
32.         finally
33.         {
34.             db. endTransaction( );    //结束事务
35.         }
36.     }
37.
38.     public void updateAge( Person person)
39.     {
40.         Log. d( AppConstants. LOG_TAG ," DBManager --> updateAge" );
41.         ContentValues cv = new ContentValues( );
42.         cv. put(" age" ,person. age);
43.         db. update( DatabaseHelper. TABLE_NAME ,cv ," name = ?" ,
44.             new String[ ]{person. name} );
45.     }
46.     public void deleteOldPerson( Person person)
47.     {
48.         Log. d( AppConstants. LOG_TAG ," DBManager --> deleteOldPerson" );
49.         db. delete( DatabaseHelper. TABLE_NAME ," age >= ?" ,
50.             new String[ ]{String. valueOf( person. age)} );
51.     }
52.     public List < Person > query( )
53.     {
54.         Log. d( AppConstants. LOG_TAG ," DBManager --> query" );
55.         ArrayList < Person > persons = new ArrayList < Person > ( );
56.         Cursor c = queryTheCursor( );
57.         while ( c. moveToNext( ))
58.         {
```

```
59.              Person person = new Person( );
60.              person. _id = c. getInt( c. getColumnIndex( "_id" ) );
61.              person. name = c. getString( c. getColumnIndex( "name" ) );
62.              person. age = c. getInt( c. getColumnIndex( "age" ) );
63.              person. info = c. getString( c. getColumnIndex( "info" ) );
64.              persons. add( person);
65.          }
66.          c. close( );
67.          return persons;
68.     }
69.     public Cursor queryTheCursor( )
70.     {
71.          Log. d( AppConstants. LOG_TAG, "DBManager --> queryTheCursor" );
72.          Cursor c = db. rawQuery( "SELECT * FROM" + DatabaseHelper. TABLE_NAME,
73.              null);
74.          return c;
75.     }
76.     public void closeDB( )
77.     {
78.          Log. d( AppConstants. LOG_TAG, "DBManager --> closeDB" );
79.          //释放数据库资源
80.          db. close( );
81.     }
82. }
```

6) 打开SQLiteOpenHelperActivity，其相应程序代码如下：

```
1.  public class SQLiteOptenHelperActivity extends Activity{
2.      private DBManager dbManager;
3.      private ListView listView;
4.      protected void onCreate( Bundle savedInstanceState)
5.      {
6.          super. onCreate( savedInstanceState);
7.          setContentView( R. layout. main);
8.          listView = ( ListView) findViewById( R. id. listView);
9.          //初始化 DBManager
10.         dbManager = new DBManager( this);
11.     }
12.
13.     public boolean onCreateOptionsMenu( Menu menu)
14.     {
15.         getMenuInflater( ). inflate( R. menu. sqlite_opten_helper, menu);
16.         return true;
```

```
17.        }
18.        protected void onDestroy()
19.        {
20.            super.onDestroy();
21.            dbManager.closeDB();//释放数据库资源
22.        }
23.        public void add(View view)
24.        {
25.            ArrayList<Person> persons = new ArrayList<Person>();
26.            Person person1 = new Person("Ella",22,"lively girl");
27.            Person person2 = new Person("Jenny",22,"beautiful girl");
28.            Person person3 = new Person("Jessica",23,"sexy girl");
29.            Person person4 = new Person("Kelly",23,"hot baby");
30.            Person person5 = new Person("Jane",25,"a pretty woman");
31.            persons.add(person1);
32.            persons.add(person2);
33.            persons.add(person3);
34.            persons.add(person4);
35.            persons.add(person5);
36.            dbManager.add(persons);
37.        }
38.        public void update(View view)
39.        {
40. //把Jane的年龄改为30(注意更改的是数据库中的值,要进行查询操作才能实现ListView的刷新)
41.            Person person = new Person();
42.            person.name = "Jane";
43.            person.age = 30;
44.            dbManager.updateAge(person);
45.        }
46.        public void delete(View view)
47.        {
48.            //删除所有30岁以上的人
49.            //同样是以查询方式才能查看更改结果
50.            Person person = new Person();
51.            person.age = 30;
52.            dbManager.deleteOldPerson(person);
53.        }
54.        public void query(View view)
55.        {
56.            List<Person> persons = dbManager.query();
57.            ArrayList<Map<String,String>> list = new ArrayList<Map<String,String>>();
58.            for(Person person:persons)
```

```
59.            }
60.            HashMap<String,String> map = new HashMap<String,String>();
61.            map.put("name",person.name);
62.            map.put("info",person.age + "years old," + person.info);
63.            list.add(map);
64.        }
65.        SimpleAdapter adapter = new SimpleAdapter(this,list,
66.                android.R.layout.simple_list_item_2,new String[]{"name",
67.                "info"},new int[]{android.R.id.text1,
68.                android.R.id.text2});
69.        listView.setAdapter(adapter);
70.    }
71.    public void queryTheCursor(View view)
72.    {
73.        Cursor c = dbManager.queryTheCursor();
74.        startManagingCursor(c);
75.        //托付给 Activity,使其根据自己的生命周期去管理 Cursor 的生命周期
76.        CursorWrapper cursorWrapper = new CursorWrapper(c)
77.        {
78.            @Override
79.            public String getString(int columnIndex)
80.            {
81.                //在简介前加上年龄
82.                if(getColumnName(columnIndex).equals("info"))
83.                {
84.                    int age = getInt(getColumnIndex("age"));
85.                    return age + "years old," + super.getString(columnIndex);
86.                }
87.                return super.getString(columnIndex);
88.            }
89.        };
90.        //确保查询结果中有"_id"列
91.        SimpleCursorAdapter adapter = new SimpleCursorAdapter(this,
92.            android.R.layout.simple_list_item_2,cursorWrapper,
93.            new String[]{"name","info"},new int[]{
94.            android.R.id.text1,android.R.id.text2});
95.        ListView listView = (ListView)findViewById(R.id.listView);
96.        listView.setAdapter(adapter);
97.    }
98. }
```

运行 chap07_5 应用,单击"添加"和"查询"按钮后,可得到如图 7-13 所示的运行界面。

上述两个案例创建了名为 TestDB.db 的数据库。这个数据库文件是存储在目录 "\data\data\你的程序报名\databases" 下，打开 DDMS 界面，如图 7-14 所示。

图 7-13　chap07_5 运行界面

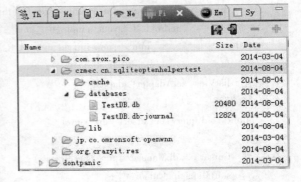

图 7-14　数据库目录

如何导出以及操作数据库呢？可以使用 SQLite3 工具。

【知识学习】 **SQLiteOpenHelper 类**

前面为读者详细讲述了使用 SQLiteDatabase 类对 SQLite 数据库进行操作的方法。但上述做法比较烦琐，即程序首先尝试向表中插入数据，如果程序抛出异常，说明被插入表的不存在，那么在异常捕获的 Catch 模块中创建这个不存在的表，然后再插入。这种做法显然烦琐，鉴于此，Android 给开发者提供了功能强大的 SQLiteOpenHelper 类来实现对数据库的操作。

SQLiteOpenHelper 类其实是 Android 系统为了方便开发者访问 SQLite 数据库而专门提供的一个工具类，如同开发者在编写 Java SE 或 Java Web 应用系统时自己编写的一个数据库操作的工具类，所不同的是，在 Android 开发中，系统已经提供了这样的工具类，开发者不用自己去编写这个工具类了。

SQLiteOpenHelper 类为开发者提供了如下方法以实现对数据库的操作。

1）synchronized SQLiteDatabase getReadableDatabase：以只读的方式打开数据库对应的 SQLiteDatabase 对象。

2）synchronized SQLiteDatabase getWritableDatabase：以读/写的方式打开数据库对应的 SQLiteDatabase 对象。

3）abstract void onCreate(SQLiteDatabase db)：当第一次创建数据库时回调该方法。开发者在创建 SQLiteOpenHelper 的子类时需要重写 onCreate 方法，这样当开发者使用方法 getReadableDatabase 或方法 getWritableDatabase 获取数据库实例失败（如数据库不存在）时，Android 系统就会自动生成一个数据库并调用 onCreate 方法。此时，开发者可通过 onCreate 方法编写生成的数据库中表的结构及添加一些数据的代码。

4）abstract void onUpgrade(SQLiteDatabase db, int oldVersion, int newVersion)：当数据库的

版本更新的时候回调该方法。开发者在创建 SQLiteOpenHelper 的子类时需要重写 onUpgrade 方法。这个方法用于升级软件时更新数据库中表的结构，当且仅当数据库的版本号发生变化时才会调用。

5）synchronized void close：关闭所有打开的 SQLiteDatabase 对象。

任务3　使用 SQLite3 工具操作数据库

【任务描述】

Android 的 tools 目录下有一个 SQLite3.exe 工具，这是一个简单的 SQLite 数据库管理工具，开发者可以使用这个工具实现对数据库的导出及简单操作。

任务 2 中两个案例创建了 1 个数据库 TestDB.db，那么如何导出及操作这两个数据库呢？可以使用 SQLite3 工具。

【任务实施】

1）将数据库 TestDB.db 从模拟器中导出到本地计算机。具体操作方法是，打开 DDMS，找到路径下的数据库文件，选中后单击"pull a file from the device"按钮即可。这里将其导出到"E:\AndroidProject\DB"目录下。

2）打开 cmd 窗口，将当前路径修改为自己的 Android 的 tools 路径下，如图 7-15 所示。

图 7-15　Android tools 安装路径

3）输入如下命令，启动 SQLite 数据库：sqlite3 e:\AndroidProject\DB\TestDB.db。启动成功后，在提示符下输入"databases"后，可以看到当前数据库，如图 7-16 所示。

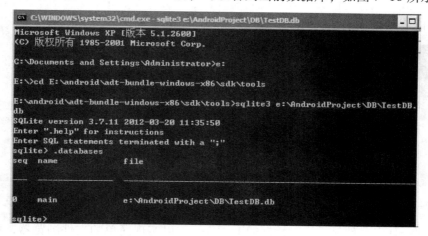

图 7-16　查看当前数据库

4）可以查询数据库中有哪些表，并输入 SQL 语句查询表中的数据，如图 7-17 所示。

需要注意的是，SQLite 内部只支持 null、integer、real（浮点数）、text（文本）和 blob（大二进制对象）5 种数据类型。但 SQLite 完全可以接受 varchar(n)、char(n)、decimal(p,s)等数据类型，SQLite 会在运算或保存时自动将它们转换为上述 5 种数据类型中相应的类型。

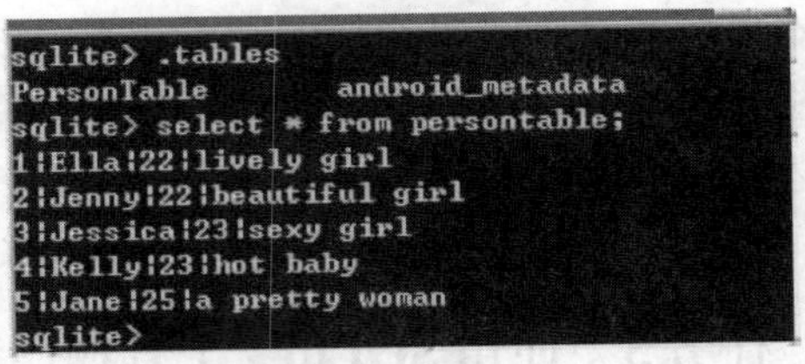

图 7-17 查看当前数据库中的表及表中的数据

【知识学习】**SQLite3 常用的命令**

SQLite3 常用的命令如下。
1）.databases：查看当前数据库。
2）.tables：查看当前数据库中的表。
3）.help：查看 SQLite3 支持的所有命令。
开发者可以使用 SQL 语句在命令窗口中操作数据库中的表。

【练习】

1. 对项目三中的习题"模仿实现 QQ 登录布局界面"继续修改，要求如下：
1）在 SQLite 数据库中创建数据库"test"，并创建一个"userinfo"数据表；
2）实现用户登录功能；
3）当用户登录成功后，显示登录成功界面，如果登录失败，显示登录失败界面；
4）需要在登录时进行用户名及密码的验证；
5）使用 SharedPerferences 实现用户勾选"记住密码"的功能；
6）将数据库从虚拟手机模拟器中导出并使用 SQLite3 工具对其操作，如对表进行增、删、改、查操作。

2. 制作一个简易的文本阅读器，以对话框实现文件浏览器，将选择的 TXT 文件内容在有 TextView 的 ScrollView 中显示。使用 SharedPreferences 保存用户的设置（字体大小及阅读模式，阅读模式可以分为白天模式与夜间模式，字体分为小、中、大 3 种），以便于用户再次使用软件时自动读取上一次的用户配置，如图 7-18 ~ 7-20 所示。

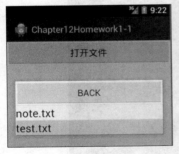
图 7-18 打开文件

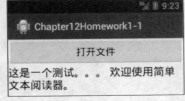
图 7-19 阅读文件

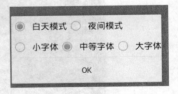
图 7-20 设定文本配置

提示：将 SD 卡文件显示到对话框的 ListView 中以实现简单的文件浏览器，使用 FileInputStream 读取文件并将其显示到 TextView 中，使用菜单项修改用户配置并保存在 SharedPreferences 中，在文本读取时设置 SharedPreferences 中的配置。

第二篇 应用篇

项目八 欧洲杯信息采集 APP 应用

本项目基于 Android 平台，采用第一篇介绍的常用 Android 技术知识点实现欧洲杯信息采集系统 APP 项目的研发与制作，这里将整个项目拆分成单独模块，系统性地实践 Android 编程在实际业务当中的运用。

【知识目标】

- Android 四大组件之一的 Activity 的相关知识点
- 灵活使用适配器，将数据集合显示到界面列表项
- 学会在 Android 系统中解析 XML 信息并将其封装成对象
- 学会使用可扩展列表及可扩展列表项与列表项的嵌套使用
- 使用并控制对话框实现业务数据的显示

【模块分解】

- 欧洲杯主视图界面的实现
- 积分榜功能的实现
- 赛事明细列表功能的实现

模块一 欧洲杯主视图界面的实现

【模块描述】

欧洲杯信息采集 APP 主要分为主界面、积分榜及赛事明细 3 个子项目。本模块主要基于线性布局 LinearLayout、主题 Theme、ImageView 组件、ListView 组件、SimpleAdapter 数据适配器、自定义布局文件等实现欧洲杯主视图界面。

本模块将实现如下操作：
1) 基于自定义布局文件实现欧洲杯主界面；
2) 自定义数据适配器将国家名和图标以列表方式呈现到主界面上。

知识点	技能点
➢ Android 常见布局管理器尤其是线性布局的用法 ➢ Android 的图片、字符串、数组等资源的引用方法 ➢ 基于 SimpleAdapter 实现自定义数据适配器的方法 ➢ 基于 Java 代码及 XML 混合搭建 UI 界面的方法 ➢ Activity 状态、生命周期及加载模式 ➢ Android 的事件处理机制	➢ XML 文件中数据的读/写操作 ➢ Android 应用程序的创建、打包、签名及运行 ➢ 用自定义数据适配器填充 ListView 组件操作

任务1 构建欧洲杯主界面

【任务描述】

基于 XML 布局文件构建欧洲杯信息采集 APP 的主界面。

【任务实施】

1）创建一个 Android 项目，取名"EuropeCup2012"。

① 在工程 res 文件夹中新建 drawable 文件夹。

② 将图 8-1 中"2012 欧洲杯"图片（见随书资源项目七中的 drawable 文件夹中）复制到当前项目的 drawable 文件夹。

③ 在工程 res 文件夹中新建 xml 文件夹。

④ 将 europe_cup.xml 文件（在随书配套资源项目七中的 xml 文件夹中）复制到当前项目的 xml 文件夹。

2）构建欧洲杯主界面。

① 主界面布局文件名为 ec2012_layout.xml。

② 主界面布局选用 LinearLayout。

③ 在主界面布局文件中增加两个 View：第一个视图为 TextView，第二个视图为 ListView，ListView 必须出现在 TextView 的下方。

④ 为 TextView 增加显示图片，显示效果如图 8-1 所示。

⑤ 将当前窗体设置为全屏，如图 8-2 所示。

图 8-1 主界面效果图（a）

图 8-2 主界面效果图（b）

创建主界面布局文件 ec2012_layout.xml 的程序代码如下所示：

```
1.   <LinearLayout xmlns:android="http://schemas.android.com/apk/res/android"
2.       android:layout_width="match_parent"
3.       android:layout_height="match_parent"
4.       android:orientation="vertical">
5.
6.       <TextView
7.           android:id="@+id/euroup_country_title"
8.           android:layout_width="wrap_content"
9.           android:layout_height="100dp"
10.          android:background="@drawable/banner"
11.          android:textAppearance="?android:attr/textAppearanceMedium"
12.          android:textColor="#00000000"/>
13.      <ListView
14.          android:id="@+id/lstCountry"
15.          android:layout_width="match_parent"
16.          android:layout_height="match_parent"
17.          android:background="#97FFFFBF"
18.          android:entries="@array/country">
19.      </ListView>
20.  </LinearLayout>
```

3）构建欧洲杯分组布局。

① 欧洲杯分组布局文件名为 europe_layout.xml。

② 主界面布局选用 LinearLayout。

③ 在主界面布局文件中增加 8 个 View：第 1～4 个视图为 TextView，第 5～8 个视图为 ImageView。一共分为两行，每行显示两个 ImageView 和两个 TextView。布局格式如图 8-3 所示。

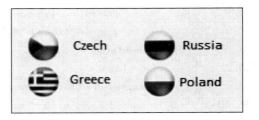

图 8-3 欧洲杯分组布局

④ 显示内容与顺序分别为：图片为 czech.png、russian.png、greece.png、poland.png；文字为 Czech、Russian、Greece、Poland。

创建欧洲杯分组布局文件 europe_layout.xml 的程序代码如下所示：

```
1.   <?xml version="1.0" encoding="utf-8"?>
2.   <LinearLayout xmlns:android="http://schemas.android.com/apk/res/android"
3.       android:layout_width="match_parent"
4.       android:layout_height="match_parent"
5.       android:gravity="center_vertical|center_horizontal"
6.       android:orientation="vertical"
7.       android:padding="5dp">
```

```
8.      < LinearLayout
9.          android:id = "@+id/euroup_country_line1"
10.         android:layout_width = "wrap_content"
11.         android:layout_height = "wrap_content"
12.         android:gravity = "center_vertical"
13.         android:orientation = "horizontal"
14.         android:layout_marginBottom = "10dp" >
15.         < LinearLayout
16.             android:id = "@+id/euroup_country_panel1"
17.             android:layout_width = "wrap_content"
18.             android:layout_height = "wrap_content"
19.             android:gravity = "center_vertical"
20.             android:orientation = "horizontal" >
21.             < ImageView
22.                 android:id = "@+id/euroup_country_icon1"
23.                 android:layout_width = "wrap_content"
24.                 android:layout_height = "wrap_content"
25.                 android:contentDescription = "@null"/>
26.             < TextView
27.                 android:id = "@+id/euroup_country_name1"
28.                 android:layout_width = "100dp"
29.                 android:layout_height = "wrap_content"
30.                 android:layout_marginLeft = "15dp"
31.                 android:textAppearance = "?android:attr/textAppearanceSmall"
32.                 android:textColor = "#00000000"/>
33.         </LinearLayout >
34.         < LinearLayout
35.             android:id = "@+id/euroup_country_panel2"
36.             android:layout_width = "wrap_content"
37.             android:layout_height = "wrap_content"
38.             android:gravity = "center_vertical"
39.             android:orientation = "horizontal" >
40.             < ImageView
41.                 android:id = "@+id/euroup_country_icon2"
42.                 android:layout_width = "wrap_content"
43.                 android:layout_height = "wrap_content"
44.                 android:contentDescription = "@null"/>
45.             < TextView
46.                 android:id = "@+id/euroup_country_name2"
47.                 android:layout_width = "60dp"
48.                 android:layout_height = "wrap_content"
49.                 android:layout_marginLeft = "10dp"
```

```
50.            android:textAppearance = "?android:attr/textAppearanceSmall"
51.            android:textColor = "#00000000"/>
52.        </LinearLayout>
53.    </LinearLayout>
54.    <LinearLayout
55.        android:id = "@+id/euroup_country_line2"
56.        android:layout_width = "wrap_content"
57.        android:layout_height = "wrap_content"
58.        android:gravity = "center_vertical"
59.        android:orientation = "horizontal" >
60.        <LinearLayout
61.            android:id = "@+id/euroup_country_panel3"
62.            android:layout_width = "wrap_content"
63.            android:layout_height = "wrap_content"
64.            android:gravity = "center_vertical"
65.            android:orientation = "horizontal" >
66.            <ImageView
67.                android:id = "@+id/euroup_country_icon3"
68.                android:layout_width = "wrap_content"
69.                android:layout_height = "wrap_content"
70.                android:contentDescription = "@null"/>
71.            <TextView
72.                android:id = "@+id/euroup_country_name3"
73.                android:layout_width = "100dp"
74.                android:layout_height = "wrap_content"
75.                android:layout_marginLeft = "15dp"
76.                android:textAppearance = "?android:attr/textAppearanceSmall"
77.                android:textColor = "#00000000"/>
78.        </LinearLayout>
79.        <LinearLayout
80.            android:id = "@+id/euroup_country_panel4"
81.            android:layout_width = "wrap_content"
82.            android:layout_height = "wrap_content"
83.            android:gravity = "center_vertical"
84.            android:orientation = "horizontal" >
85.            <ImageView
86.                android:id = "@+id/euroup_country_icon4"
87.                android:layout_width = "wrap_content"
88.                android:layout_height = "wrap_content"
89.                android:contentDescription = "@null"/>
90.            <TextView
91.                android:id = "@+id/euroup_country_name4"
```

```
92.                    android:layout_width = "60dp"
93.                    android:layout_height = "wrap_content"
94.                    android:layout_marginLeft = "10dp"
95.                    android:textAppearance = "?android:attr/textAppearanceSmall"
96.                    android:textColor = "#00000000"/>
97.            </LinearLayout>
98.        </LinearLayout>
99.    </LinearLayout>
```

任务2 列表呈现国家名和图标

【任务描述】

本任务是显示完整的欧洲杯分组界面（使用 ListView 呈现）。本任务的功能实现需要运用 XML 数据源，因此首先应该实现数据源文件 europe_cup.xml 和相关的实体类，然后创建 EuropeCup2012Manager 组件，用来读取 res/xml/europe_cup.xml 文件中的所有数据，并将其保存在 EuropeCup2012Score 对象中，该对象可以通过 getScore 方法返回。接下来再创建 SimpleAdapter 组件，将 SimpleAdapter 对象设置于 ec2012_layout.xml 中的 ListView 视图。

【任务实施】

1）创建实体类 Team、Player、Home、Group、Game、Away、EuropeCup2012Score 等。com.frank.ec2012.entity 包下所有的实体类的关系图如图 8-4 所示。

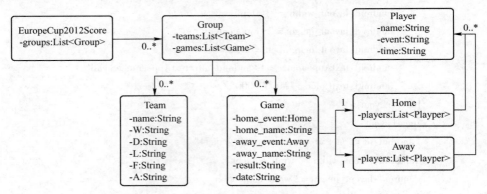

图 8-4 实体类关系图

其中，Team 相关程序代码如下所示：

```
1.   public class Team{
2.       private String name = null;
3.       private String W = null;
4.       private String D = null;
5.       private String L = null;
6.       private String F = null;
```

```
7.      private String A = null;
8.      private String pts = null;
9.      public String getName() {
10.         return name;
11.     }
12.     public void setName(String name) {
13.         this.name = name;
14.     }
15.     ...//省略其他的set/get方法
16. }
```

Game 实体类相关程序代码如下所示:

```
1.  public class Game{
2.      private Home home_event = null;
3.      private Away away_event = null;
4.      private String home_team = null;
5.      private String away_team = null;
6.      private String result = null;
7.      private String date = null;
8.      public Home getHome_event() {
9.          return home_event;
10.     }
11.     public void setHome_event(Home home_event) {
12.         this.home_event = home_event;
13.     }
14.     public Away getAway_event() {
15.         return away_event;
16.     }
17.     public void setAway_event(Away away_event) {
18.         this.away_event = away_event;
19.     }
20.     ...//省略其他的set/get方法
21. }
```

2) 创建 XML 数据源 europe_cup.xml, 用来保存相关大赛数据, 其相关程序代码如下所示:

```
1.  <europe_cup>
2.      <group name="A">
3.          <teams>
4.              <team name="Czech" W="2" D="0" L="1" F="4" A="5" Pts="6"/>
5.              <team name="Greece" W="1" D="1" L="1" F="3" A="3" Pts="4"/>
6.              <team name="Russian" W="1" D="1" L="1" F="5" A="3" Pts="4"/>
```

```
7.                   < team name = " Poland" W = "0" D = "2" L = "1" F = "2" A = "3" Pts = "2" / >
8.               </ teams >
9.               < games >
10.                  < game home = " Poland" away = " Greece" result = "1 : 1" date = "8 June 2012" >
11.                      < home >
12.                          < player name = " Lewandowski" event = " goals" time = "17" / >
13.                          < player name = " Szczesny" event = " reds" time = "69" / >
14.                  < player name = " Rybus" event = " substs" time = "70" / >
15.                  < player name = " Tyton" event = " substs" time = "70" / >
16.                      </ home >
17.                      < away >
18.                          < player name = " Papastathopoulos" event = " yells" time = "35" / >
19.                          < player name = " Papadopoulos" event = " substs" time = "37" / >
20.                      < player name = " Unkonwplayer" event = " substs" time = "37" / >
21.                      < player name = " Papastathopoulos" event = " yell_reds" time = "44" / >
22.                      < player name = " Jose Holebas" event = " yells" time = "45 + 2" / >
23.                      < player name = " Sotiris Ninis" event = " substs" time = "46" / >
24.                      < player name = " Salpingidis" event = " substs" time = "46" / >
25.                      < player name = " Salpingidis" event = " goals" time = "51" / >
26.                      < player name = " Karagounis" event = " yells" time = "54" / >
27.                      < player name = " Fanis Gekas" event = " substs" time = "68" / >
28.                      < player name = " Kostas Fortounis" event = " substs" time = "68" / >
29.                      </ away >
30.                  </ game >
31.                  < games >
                                       ⋮
109. </ group >
110.    < group name = " B" >
111.        < teams >
112.            < team name = " Germany" W = "2" D = "0" L = "0" F = "3" A = "1" Pts = "6" / >
113.            < team name = " Portugal" W = "1" D = "0" L = "1" F = "3" A = "3" Pts = "3" / >
114.            < team name = " Denmark" W = "1" D = "0" L = "1" F = "3" A = "3" Pts = "3" / >
115.            < team name = " Netherlands" W = "0" D = "0" L = "2" F = "1" A = "3" Pts = "0" / >
116.        </ teams >
117.        < games >
118.            < game home = " Netherlands" away = " Denmark" result = "0 : 1" date = "9 June 2012" >
119.                < sub >
120.                    < home >
121.                        < player name = " van Bommel" event = " yells" time = "67" / >
122.                        < player name = " de Jong" event = " substs" time = "71" / >
123.                        < player name = " Huntelaar" event = " substs" time = "71" / >
124.                        < player name = " Afellay" event = " yells" time = "71" / >
```

```
125.                    < player name = " van der Vaart" event = " substs" time = "71"/ >
126.                    < player name = " van der Wiel" event = " substs" time = "85"/ >
127.                    < player name = " Dirk Kuyt" event = " substs" time = "85"/ >
128.                </home >
129.                < away >
130.                    < player name = " Krohn – Dehli" event = " goals" time = "24"/ >
131.                    < player name = " Eriksen" event = " substs" time = "74"/ >
132.                    < player name = " Lasse Schone" event = " substs" time = "74"/ >
133.                    < player name = " Simon Poulsen" event = " yells" time = "78"/ >
134.                    < player name = " William Kvist" event = " yells" time = "81"/ >
135.                    < player name = " Rommedahl" event = " substs" time = "84"/ >
136.                    < player name = " Mikkelsen" event = " substs" time = "84"/ >
137.                </away >
138.              </sub >
139.            </game >
                         ⋮
218.          </games >
219.       </group >
220.       < group name = "C" >
221.          < teams >
222.             < team name = "Spain" W = "1" D = "1" L = "0" F = "5" A = "1" Pts = "4"/ >
223.             < team name = "Croatia" W = "1" D = "1" L = "0" F = "4" A = "2" Pts = "4"/ >
224.             < team name = "Italy" W = "0" D = "2" L = "0" F = "2" A = "2" Pts = "2"/ >
225.             < team name = "Ireland" W = "0" D = "0" L = "2" F = "1" A = "7" Pts = "0"/ >
226.          </teams >
227.          < games >
228.             < game home = "Spain" away = "Italy" result = "1:1" date = "10 June 2012" >
229.                < sub >
230.                   < home >
231.                      < player name = " Cesc Fabregas" event = " goals" time = "64"/ >
232.                      < player name = " David Silva" event = " substs" time = "64"/ >
233.                      < player name = " Jesus Navas" event = " substs" time = "64"/ >
234.                      < player name = " Cesc Fabregas" event = " substs" time = "74"/ >
235.                      < player name = " Fernando Torres" event = " substs" time = "74"/ >
236.                      < player name = " Alvaro Arbeloa" event = " yells" time = "84"/ >
237.                      < player name = " Fernando Torres" event = " yells" time = "84"/ >
238.                   </home >
239.                   < away >
240.                      < player name = " Mario Balotelli" event = " yells" time = "37"/ >
241.                      < player name = " Mario Balotelli" event = " substs" time = "56"/ >
242.                      < player name = " Antonio Di Natale" event = " substs" time = "56"/ >
243.                      < player name = " Antonio Cassano" event = " substs" time = "65"/ >
```

```
244.                    < player name = " Sebastian Giovinco" event = " substs" time = "65" / >
245.                    < player name = " Leonardo Bonucci" event = " yells" time = "66" / >
246.                    < player name = " Giorgio Chiellini" event = " yells" time = "79" / >
247.                    < player name = " Christian Maggio" event = " yells" time = "89" / >
248.                    < player name = " Thiago Motta" event = " substs" time = "90" / >
249.                    < player name = " Antonio Nocerino" event = " substs" time = "90" / >
250.                </ away >
251.              </ sub >
252.            </ game >
                  ⋮
334.          </ games >
335.        </ group >
336.        < group name = " D" >
337.          < teams >
338.            < team name = " Ukraine" W = "1" D = "0" L = "0" F = "2" A = "1" Pts = "3" / >
339.            < team name = " England" W = "0" D = "1" L = "0" F = "1" A = "0" Pts = "1" / >
340.            < team name = " France" W = "0" D = "1" L = "0" F = "1" A = "0" Pts = "1" / >
341.            < team name = " Sweden" W = "0" D = "0" L = "1" F = "1" A = "2" Pts = "0" / >
342.          </ teams >
343.          < games >
344.            < game home = " France" away = " England" result = " 1 : 1" date = " 11 June 2012" >
345.              < sub >
346.                < home >
347.                    < player name = " Samir Nasri" event = " goals" time = "39" / >
348.                    < player name = " Florent Malouda" event = " substs" time = "85" / >
349.                    < player name = " Marvin Martin" event = " substs" time = "85" / >
350.                    < player name = " Yohan Cabaye" event = " substs" time = "85" / >
351.                    < player name = " Hatem Ben Arfa" event = " substs" time = "85" / >
352.                </ home >
353.                < away >
354.                    < player name = " Joleon Lescott" event = " goals" time = "36" / >
355.                    < player name = " Alex Chamberlain" event = " yells" time = "34" / >
356.                    < player name = " Ashley Young" event = " yells" time = "71" / >
357.                    < player name = " Alex Chamberlain" event = " substs" time = "77" / >
358.                    < player name = " Jermain Defoe" event = " substs" time = "77" / >
359.                    < player name = " Scott Parker" event = " substs" time = "78" / >
360.                    < player name = " Jordan Henderson" event = " substs" time = "78" / >
361.                    < player name = " Danny Welbeck" event = " substs" time = "90" / >
362.                    < player name = " PTheo Walcott" event = " substs" time = "90" / >
363.                </ away >
364.              </ sub >
365.            </ game >
                  ⋮
```

```
391.        </games>
392.      </group>
393.  </europe_cup>
```

3)创建 EuropeCup2012Manager 组件,用于读取 europe_cup.xml 文件中的所有数据,其相应程序代码如下:

```
1.   public class EuropeCup2012Manager{
2.       private static EuropeCup2012Manager instance = null;
3.       private EuropeCup2012Score score = null;
4.       public static String ROOT_NODE_NAME = "europe_cup";
5.       public static String GROUP_NODE_NAME = "group";
6.       public static String TEAM_NODE_NAME = "team";
7.       public static String GAME_NODE_NAME = "game";
8.       public static String HOME_NODE_NAME = "home";
9.       public static String AWAY_NODE_NAME = "away";
10.      public static String PLAYER_NODE_NAME = "player";
11.      private XmlResourceParser parser = null;
12.      private EuropeCup2012Manager(Context context){
13.          this.parser = context.getResources().getXml(R.xml.europe_cup);
14.      }
15.      public EuropeCup2012Score getScore(){
16.          if(score != null)
17.              return score;
18.          boolean isHome_Event = true;
19.          try{
20.              while (parser.getEventType() != XmlResourceParser.END_DOCUMENT){
21.                  if (parser.getEventType() != XmlResourceParser.START_TAG){
22.                      parser.next();
23.                      continue;
24.                  }
25.                  String tagName = parser.getName();
26.                  if (ROOT_NODE_NAME.equals(tagName)){
27.                      score = new EuropeCup2012Score();
28.                      List<Group> groups = new ArrayList<Group>();
29.                      score.setGroups(groups);
30.                  }
31.                  if (GROUP_NODE_NAME.equals(tagName)){
32.                      Group group = createGroup();
33.                      score.getGroups().add(group);
34.                  }
35.                  if (TEAM_NODE_NAME.equals(tagName)){
36.                      Team team = createTeam();
37.                      score.getLastGroup().getTeams().add(team);
38.                  }
```

```java
39.             if (GAME_NODE_NAME.equals(tagName)){
40.                 Game game = createGame();
41.                 score.getLastGroup().getGames().add(game);
42.             }
43.             if (HOME_NODE_NAME.equals(tagName)){
44.                 Home home = createHome();
45.                 score.getLastGroup().getLastGame().setHome_event(home);
46.                 isHome_Event = true;
47.             }
48.             if (AWAY_NODE_NAME.equals(tagName)){
49.                 Away away = createAway();
50.                 score.getLastGroup().getLastGame().setAway_event(away);
51.                 isHome_Event = false;
52.             }
53.             if (PLAYER_NODE_NAME.equals(tagName)){
54.                 Player player = createPlayer();
55.                 if (isHome_Event)
56.                     score.getLastGroup().getLastGame().getHome_event()
57.                         .getPlayers().add(player);
58.                 else
59.                     score.getLastGroup().getLastGame().getAway_event()
60.                         .getPlayers().add(player);
61.             }
62.             parser.next();
63.         }
64.     }catch(XmlPullParserException e){
65.         return null;
66.     }catch(IOException e){
67.         return null;
68.     }
69.     return score;
70. }
71. private Player createPlayer(){
72.     Player player = new Player();
73.     player.setName(this.parser.getAttributeValue(null,"name"));
74.     player.setEvent(this.parser.getAttributeValue(null,"event"));
75.     player.setTime(this.parser.getAttributeValue(null,"time"));
76.     return player;
77. }
78. private Home createHome(){
79.     Home home = new Home();
80.     home.setPlayers(new ArrayList<Player>());
81.     return home;
82. }
83. private Away createAway(){
```

```java
84.            Away away = new Away();
85.            away.setPlayers(new ArrayList<Player>());
86.            return away;
87.        }
88.        private Game createGame(){
89.            Game game = new Game();
90.            game.setHome_team(this.parser.getAttributeValue(null,"home"));
91.            game.setAway_team(this.parser.getAttributeValue(null,"away"));
92.            game.setResult(this.parser.getAttributeValue(null,"result"));
93.            game.setDate(this.parser.getAttributeValue(null,"date"));
94.            return game;
95.        }
96.        private Team createTeam(){
97.            Team team = new Team();
98.            team.setA(this.parser.getAttributeValue(null,"A"));
99.            team.setD(this.parser.getAttributeValue(null,"D"));
100.           team.setL(this.parser.getAttributeValue(null,"L"));
101.           team.setW(this.parser.getAttributeValue(null,"W"));
102.           team.setName(this.parser.getAttributeValue(null,"name"));
103.           team.setF(this.parser.getAttributeValue(null,"F"));
104.           team.setPts(this.parser.getAttributeValue(null,"Pts"));
105.           return team;
106.       }
107.       private Group createGroup(){
108.           Group group = new Group();
109.           group.setName(this.parser.getAttributeValue(null,"name"));
110.           group.setGames(new ArrayList<Game>());
111.           group.setTeams(new ArrayList<Team>());
112.           return group;
113.       }
114.       public static EuropeCup2012Manager getInstance(Context context){
115.           if(instance == null)
116.               instance = new EuropeCup2012Manager(context);
117.           return instance;
118.       }
119.   }
```

上述程序代码中 EuropeCup2012Manager 组件用于读取 res/xml/europe_cup.xml 文件中的所有数据，并保存在 EuropeCup2012Score 对象中，该对象可以通过 getScore 方法返回。每个实体类名与类结构都和 res/xml/europe_cup.xml 文件中的节点名对应。获取 EuropeCup2012Score 对象等同于获取整个 XML 的数据。下面给出通过实体类提供属性的方式获取 XML 节点数据的步骤。

① 保证加载 XML 文件的操作永远只做一次。XML 文件读取类 EuropeCup2012Manager 时需要被构建成单例模式具体步骤如下：
　　a. 将 EuropeCup2012Manager 的构造函数设置为私有；
　　b. 将 getInstance 方法设置为静态；
　　c. 在 EuropeCup2012Manager 类中设置静态的 EuropeCup2012Manager 对象；
　　d. 判断 EuropeCup2012Manager 对象是否为 null。如果是，则创建一个新对象，否则返回当前 EuropeCup2012Manager 对象；
　　e. 在私有构造函数中创建 XMLResourceParser；
　　f. 在私有构造函数中读取所有 XML 数据。
② 将数据保存入固定的实体对象中的具体步骤如下。
　　a. 处理 europe_cup 节点。如果当前解析器解析到了 europe_cup 节点的起始位置，那么开始处理：创建 EuropeCup2012Score 对象；创建 Group 集合对象；将新建的 Group 集合对象设置给 EuropeCup2012Score。
　　b. 处理 Group 节点。如果当前解析器解析到了 Group 节点的起始位置，那么开始处理：创建 Group 对象；将新建的 Group 对象设置给 EuropeCup2012Score 内部的 Group 集合对象；读取 Group 节点中的 name 属性，并将其设置给 Group 对象的 name 属性；新建 Team 集合，将新建的 Team 集合对象设置给 Group；新建 Game 集合，将新建的 Game 集合对象设置给 Group。
　　c. 处理 Team 节点。如果当前解析器解析到了 Team 节点的起始位置，那么开始处理：创建 Team 对象；将新建的 Team 对象设置给 EuropeCup2012Score 内部 Group 集合中最后一个 Group 对象的 Team 集合对象；读取 Team 节点中的 A 属性，并将其设置给 Team 对象的 A 属性；读取 Team 节点中的 D 属性，并将其设置给 Team 对象的 D 属性；读取 Team 节点中的 L 属性，并将其设置给 Team 对象的 L 属性；读取 Team 节点中的 W 属性，并将其设置给 Team 对象的 W 属性；读取 Team 节点中的 F 属性，并将其设置给 Team 对象的 F 属性；读取 Team 节点中的 name 属性，并将其设置给 Team 对象的 name 属性；读取 Team 节点中的 Pts 属性，并将其设置给 Team 对象的 Pts 属性。
　　d. 处理 Game 节点。如果当前解析器解析到了 Game 节点的起始位置，那么开始处理：创建 Game 对象；将新建的 Game 对象设置给 EuropeCup2012Score 内部 Group 集合中最后一个 Group 对象的 Game 集合对象；读取 Game 节点中的 home 属性，并将其设置给 Game 对象的 home_team 属性；读取 Game 节点中的 away 属性，并将其设置给 Game 对象的 away_team 属性；读取 Game 节点中的 result 属性，并将其设置给 Game 对象的 result 属性；读取 Game 节点中的 date 属性，并将其设置给 Game 对象的 date 属性。
　　e. 处理 Home 节点。如果当前解析器解析到了 Home 节点的起始位置，那么开始处理：创建 Home 对象；新建 Player 集合，将新建的 Player 集合对象设置给 Home 对象；将新建的 Home 对象设置给 EuropeCup2012Score 内部 Group 集合中最后一个 Group 对象中 Game 集合中最后一个 Game 的 home_event 属性。
　　f. 处理 Away 节点。如果当前解析器解析到了 Away 节点的起始位置，那么开始处理：创建 Away 对象；新建 Player 集合，将新建的 Player 集合对象设置给 Away 对象；将新建的 Away 对象设置给 EuropeCup2012Score 内部 Group 集合中最后一个 Group 对象中 Game 集合中最后一个 Game 的 away_event 属性。

g. 处理 Player 节点。如果当前解析器解析到了 Player 节点的起始位置，那么开始处理：创建 Player 对象；读取 Player 节点中的 name 属性，并将其设置给 Player 对象的 name 属性；读取 Player 节点中的 event 属性，并将其设置给 Player 对象的 event 属性；读取 Player 节点中的 time 属性，并将其设置给 Player 对象的 time 属性。

h. 判断当前解析的 Player 上级节点名。

Home：将 Player 设置给 EuropeCup2012Score 内部的 Group 集合中最后一个 Group 对象中 Game 集合中最后一个 Game 的 Home 对象的 Player 集合对象。

Away：将 Player 设置给 EuropeCup2012Score 内部 Group 集合中最后一个 Group 对象中 Game 集合中最后一个 Game 的 Away 对象的 Player 集合对象。

4）创建 CountryAdapter 数据适配器，其相应程序代码如下所示：

```
1.  public class CountryAdapter extends SimpleAdapter{
2.      private Context context = null;
3.      public CountryAdapter(Context context,List<? extends Map<String,?>> data,
4.              int resource,String[] from,int[] to){
5.          super(context,data,resource,from,to);
6.          this.context = context;
7.      }
8.      public void setViewImage(ImageView v,String value){
9.          String name = value.substring(0,1).toLowerCase() + value.substring(1);
10.         int iconId = context.getResources().getIdentifier(name,"drawable",
11.             context.getPackageName());
12.         v.setImageResource(iconId);
13.     }
14. }
```

5）打开 EuropeCup2012Activity，修改代码，加载 XML 页面，并创建 SimpleAdapter 数据适配器组件，将 SimpleAdapter 对象设置给 ec2012_layout.xml 中的 ListView 视图。其相应程序代码如下所示：

```
1.  public class EuropeCup2012Activity extends Activity{
2.      protected void onCreate(Bundle savedInstanceState){
3.          super.onCreate(savedInstanceState);
4.          super.setContentView(R.layout.ec2012_layout);
5.          initComponent();
6.      }
7.      private void initComponent(){
8.          initTeamsSequence();
9.      }
10.     private void initTeamsSequence(){
11.         ListView view = (ListView)super.findViewById(R.id.lstCountry);
```

```
12.        String[ ] counties = super. getResources( )
13.                . getStringArray( R. array. country) ;
14.        List < Map < String, Object >> parameters = new ArrayList < Map < String, Object >> ( ) ;
15.        for ( int i = 0 ; i < counties. length ; i += 4 ) {
16.            Map < String, Object > param = new HashMap < String, Object > ( ) ;
17.            for ( int j = 0 ; j < 4 ; j ++ ) {
18.                param. put( "icon" + ( j + 1) , counties[ j + i ]) ;
19.                param. put( "name" + ( j + 1) , counties[ j + i ]) ;
20.            }
21.            parameters. add( param) ;
22.        }
23.        CountryAdapter adapter = new CountryAdapter( this, parameters,
24.                R. layout. europe_layout, new String[ ] { "icon1" , "name1" ,
25.                "icon2" , "name2" , "icon3" , "name3" , "icon4" , "name4" } ,
26.                new int[ ] { R. id. euroup_country_icon1 ,
27.                    R. id. euroup_country_name1 , R. id. euroup_country_icon2 ,
28.                    R. id. euroup_country_name2 , R. id. euroup_country_icon3 ,
29.                    R. id. euroup_country_name3 , R. id. euroup_country_icon4 ,
30.                    R. id. euroup_country_name4 } ) ;
31.        view. setAdapter( adapter) ;
32.    }
33. }
```

运行该应用，得到的界面如图 8-5 所示。

图 8-5　欧洲杯信息采集 APP 运行主界面

模块二　积分榜的实现

【模块描述】

本任务是根据用户在主界面上选择的小组而呈现小组名（利用 AlertDialog）。即当用户单击图 8-5 所示的欧洲杯信息采集 APP 运行主界面中的任意小组时，将会弹出一个对话框，显示选中的小组中各个国家的比赛积分信息。

本模块将实现如下操作：
1）自定义对话框 AlertDialog 及弹出；
2）在对话框中以列表的形式显示小组内国家的积分信息。

知识点	技能点
➢ Android 的 ListView 组件事件处理机制 ➢ 基于 AlertDialog 的自定义对话框的定义 ➢ Android 数据适配器的类型及用法	➢ ListView 组件的 OnItemClickListener 事件监听器操作 ➢ 自定义对话框 AlertDialog ➢ 数据适配器的定义及使用

任务 1　积分榜对话框的创建及弹出

【任务描述】

欧洲杯信息采集 APP 中积分榜对话框的创建及弹出。

【任务实施】

1）将主界面的布局文件重新切换回 ec2012_layout.xml。
2）为 ec2012_layout.xml 中的 ListView 增加事件（OnItemClickListener）。
3）在对 ListView 的列表项单击事件中增减构建对话框的代码，其方法有如下 3 种：
① 通过 OnItemClickListener 事件的 position 了解用户单击的行号。
② 从 EuropeCup2012Manager 对象中获取 position 所对应的小组名称。
③ 将对话框的 Title 设置为：Welcome to Europe Cup 2012，建议：通过资源文件加载该 Title 内容。
4）为对话框设置一个"正面类型"的按钮，当按钮被单击后，对话框退出，其相应程序代码如下所示：

```
1.    private void initDialog( ) {
2.        ListView view = (ListView) super.findViewById( R. id. lstCountry);
3.        view. setOnItemClickListener( new OnItemClickListener( ) {
4.            public void onItemClick( AdapterView < ? > parent, View view,
5.                int position, long id) {
```

```
6.             AlertDialog dialog = createDialog(position);
7.             dialog.show();
8.         }
9.     });
10. }
11.
12. private AlertDialog createDialog(int position) {
13.     Builder dialogContentBuilder = new AlertDialog.Builder(this);
14.     /******************* Dialog Content *********************/
15.             ************部分代码省略
16.     /******************* Dialog Title **********************/
17.     AlertDialog dialog = dialogContentBuilder.create();
18.     dialog.setTitle("Welcome to Euroup Cup 2012");
19.     dialog.setIcon(R.drawable.ic_menu_largetiles);
20.
21.     /******************* Dialog Bottom *********************/
22.     dialog.setButton(DialogInterface.BUTTON_POSITIVE,"OK",
23.             new android.content.DialogInterface.OnClickListener() {
24.                 public void onClick(DialogInterface dialog, int which) {
25.                     dialog.dismiss();
26.                 }
27.
28.             });
29.
30.     return dialog;
31. }
```

得到的效果图如图8-6所示。

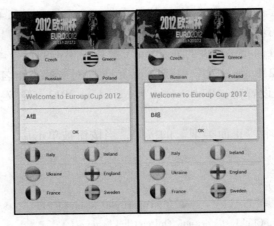

图8-6 积分榜对话框

任务 2　列表显示小组内国家积分情况

【任务描述】

继续对任务 1 进行优化,在弹出的积分榜对话框中显示小组内国家的积分信息。

【任务实施】

1）在 Adapter 包中创建一个 TeamHolder 类,用来封装小组内国家的相关组件参数,其相应程序代码如下所示:

```
1.   public class TeamHolder {
2.
3.       public TextView sequence = null;
4.       public ImageView name = null;
5.       public TextView W = null;
6.       public TextView D = null;
7.       public TextView L = null;
8.       public TextView F = null;
9.       public TextView A = null;
10.      public TextView Pts = null;
11.  }
```

2）创建小组积分榜对话框界面布局 scoreboard_layout.xml,积分榜 Title 需要设置 8 个视图对象:

① TextView:视图内容→Seq;
② TextView:视图内容→T;
③ TextView:视图内容→W;
④ TextView:视图内容→D;
⑤ TextView:视图内容→L;
⑥ TextView:视图内容→F;
⑦ TextView:视图内容→A;
⑧ TextView:视图内容→Pts。

建议将所有视图内容定义在资源文件中,其相应程序代码如下所示:

```
1.   <?xml version = "1.0" encoding = "utf-8"?>
2.   <LinearLayout xmlns:android = "http://schemas.android.com/apk/res/android"
3.       android:id = "@+id/scorelinepanel"
4.       android:layout_width = "match_parent"
5.       android:layout_height = "match_parent"
6.       android:gravity = "left|center_vertical"
7.       android:orientation = "horizontal"
8.       android:padding = "3dp" >
```

```
9.
10.        <TextView
11.            android:id="@+id/txtScoreSequence"
12.            android:layout_width="wrap_content"
13.            android:layout_height="wrap_content"
14.            android:layout_marginLeft="30dp"
15.            android:layout_marginRight="10dp"
16.            android:background="@drawable/seq"
17.            android:gravity="center_vertical|center_horizontal"
18.            android:textColor="#000000"
19.            android:textSize="12sp" />
20.        <ImageView
21.            android:id="@+id/txt_scorebar_name"
22.            android:layout_width="20dp"
23.            android:layout_height="20dp"
24.            android:contentDescription="@null"
25.            android:layout_marginRight="10dp" />
26.        <TextView
27.            android:id="@+id/txt_scorebar_w"
28.            android:layout_width="25dp"
29.            android:layout_height="wrap_content"
30.            android:layout_marginRight="3dp" />
31.        <TextView
32.            android:id="@+id/txt_scorebar_d"
33.            android:layout_width="25dp"
34.            android:layout_height="wrap_content"
35.            android:layout_marginRight="3dp" />
36.        <TextView
37.            android:id="@+id/txt_scorebar_l"
38.            android:layout_width="25dp"
39.            android:layout_height="wrap_content"
40.            android:layout_marginRight="3dp" />
41.        <TextView
42.            android:id="@+id/txt_scorebar_goal"
43.            android:layout_width="25dp"
44.            android:layout_height="wrap_content"
45.            android:layout_marginRight="3dp" />
46.        <TextView
47.            android:id="@+id/txt_scorebar_lgoal"
48.            android:layout_width="25dp"
49.            android:layout_height="wrap_content"
50.            android:layout_marginRight="3dp" />
```

```
51.        <TextView
52.            android:id = "@+id/txt_scorebar_score"
53.            android:layout_width = "25dp"
54.            android:layout_height = "wrap_content"
55.            android:layout_marginRight = "3dp" />
56.    </LinearLayout>
```

3）创建一个行管理器接口 AdapterRowManager，其中定义两个抽象方法，即 Object getRow(View convertView)和 Object getHolder(View convertView)，前者用于返回选中的行对象，后者用于获取选中行所对应的 TeamHolder 对象。

4）创建分数行管理器 ScoreAdapterRowManager，实现行管理器接口 AdapterRowManager，其相应程序代码如下：

```
1.   public class ScoreAdapterRowManager implements AdapterRowManager {
2.       private LayoutInflater inflater = null;
3.       private LinearLayout row = null;
4.       public ScoreAdapterRowManager(LayoutInflater inflater) {
5.           this.inflater = inflater;
6.       }
7.       private LinearLayout getGroupRow(View convertView) {
8.           if (convertView == null) {
9.               row = (LinearLayout) inflater.inflate(R.layout.scoreboard_layout,
10.                      null);
11.          } else {
12.              row = (LinearLayout) convertView;
13.          }
14.          return row;
15.      }
16.      private TeamHolder getTeamHolderAndSaveIt(View convertView) {
17.
18.          if(convertView != null) {
19.              return (TeamHolder)row.getTag();
20.          }
21.          TeamHolder holder = new TeamHolder();
22.          row.setTag(holder);
23.          holder.sequence = (TextView) row.findViewById(R.id.txtScoreSequence);
24.          holder.name = (ImageView) row.findViewById(R.id.txt_scorebar_name);
25.          holder.W = (TextView) row.findViewById(R.id.txt_scorebar_w);
26.          holder.D = (TextView) row.findViewById(R.id.txt_scorebar_d);
27.          holder.L = (TextView) row.findViewById(R.id.txt_scorebar_l);
28.          holder.F = (TextView) row.findViewById(R.id.txt_scorebar_goal);
29.          holder.A = (TextView) row.findViewById(R.id.txt_scorebar_lgoal);
```

```
30.            holder.Pts = (TextView) row.findViewById(R.id.txt_scorebar_score);
31.            return holder;
32.        }
33.        public Object getRow(View convertView) {
34.            return getGroupRow(convertView);
35.        }
36.        public Object getHolder(View convertView) {
37.            return getTeamHolderAndSaveIt(convertView);
38.        }
39.    }
```

上述程序代码的第 9 行是利用 LayoutInflater 对象的 inflate 方法加载 scoreboard_layout.xml。

5) 创建分数显示面板的数据适配器 ScoreboardAdapter, 用于填充积分界面, 其相应程序代码如下所示:

```
1.  public class ScoreboardAdapter extends BaseExpandableListAdapter {
2.      private Group group = null;
3.      private LayoutInflater inflater = null;
4.      private Context context = null;
5.      public ScoreboardAdapter(Group group, Context context) {
6.          this.group = group;
7.          this.context = context;
8.          inflater = (LayoutInflater) context
9.                  .getSystemService(Context.LAYOUT_INFLATER_SERVICE);
10.     }
11.     public int getGroupCount() {
12.         if (group == null)
13.             return 0;
14.         return group.getTeams().size();
15.     }
16.     public int getChildrenCount(int groupPosition) {
17.         String teamName = group.getTeams().get(groupPosition).getName();
18.         List<Game> games = this.group.findGamesByAssignedTeam(teamName);
19.         if (games == null)
20.             return 0;
21.         return games.size();
22.     }
23.     public Object getGroup(int groupPosition) {
24.         if (group == null || group.getTeams().size() <= 0)
25.             return null;
26.         return group.getTeams().get(groupPosition);
```

```
27.      }
28.      public Object getChild(int groupPosition, int childPosition) {
29.          String teamName = group.getTeams().get(groupPosition).getName();
30.          List<Game> games = this.group.findGamesByAssignedTeam(teamName);
31.          return games.get(childPosition);
32.      }
33.      public long getGroupId(int groupPosition) {
34.          return groupPosition;
35.      }
36.      public long getChildId(int groupPosition, int childPosition) {
37.          return childPosition;
38.      }
39.      public boolean hasStableIds() {
40.          return true;
41.      }
42.      public boolean isChildSelectable(int groupPosition, int childPosition) {
43.          return true;
44.      }
45.      /**
46.       * 获取主视图对象
47.       */
48.      public View getGroupView(int groupPosition, boolean isExpanded,
49.              View convertView, ViewGroup parent) {
50.          if (group == null || group.getTeams().size() <= 0)
51.              return convertView;
52.          ScoreAdapterRowManager manager = new ScoreAdapterRowManager(inflater);
53.          LinearLayout row = (LinearLayout) manager.getRow(convertView);
54.          TeamHolder holder = (TeamHolder) manager.getHolder(convertView);
55.          setDataForGroupRow(holder, groupPosition);
56.          return row;
57.      }
58.      ...
59.      private void setDataForGroupRow(TeamHolder holder, int groupPosition) {
60.          Team team = group.getTeams().get(groupPosition);
61.          holder.sequence.setText(String.valueOf(groupPosition + 1));
62.          holder.name.setImageResource(getImageIdentitifier(team.getName()));
63.          holder.W.setText(team.getW());
64.          holder.D.setText(team.getD());
65.          holder.L.setText(team.getL());
66.          holder.F.setText(team.getF());
67.          holder.A.setText(team.getA());
68.          holder.Pts.setText(team.getPts());
```

```
69.      }
70.      private int getImageIdentitifier(String imageName) {
71.          String name = imageName.substring(0,1).toLowerCase()
72.                  + imageName.substring(1);
73.          int identitifier = this.context.getResources().getIdentifier(name,
74.                  "drawable",context.getPackageName());
75.          return identitifier;
76.      }
77. }
```

6) 打开 EuropeCup2012Activity, 修改其中的方法 createDialog, 其相应的程序代码如下:

```
1.  private AlertDialog createDialog(int position) {
2.      Builder dialogContentBuilder = new AlertDialog.Builder(this);
3.      /******************* Dialog Content *********************/
4.      LinearLayout dialogContentLayout = new LinearLayout(this);
5.      dialogContentLayout.setLayoutParams(new LayoutParams(
6.              LayoutParams.MATCH_PARENT,LayoutParams.MATCH_PARENT));
7.      dialogContentLayout.setOrientation(LinearLayout.VERTICAL);
8.      LayoutInflater inflater =
9.              (LayoutInflater)getSystemService(Activity.LAYOUT_INFLATER_SERVICE);
10.     dialogContentLayout.addView(inflater.inflate(R.layout.scoreboard_title_layout,null));
11.     ExpandableListView elv = new ExpandableListView(this);
12.     Group group = EuropeCup2012Manager.getInstance(this).getScore()
13.             .getGroup(position);
14.     ScoreboardAdapter adapter = new ScoreboardAdapter(group,this);
15.     elv.setAdapter(adapter);
16.     elv.setScrollbarFadingEnabled(false);
17.     dialogContentLayout.addView(elv);
18.     dialogContentBuilder.setView(dialogContentLayout);
19.     /******************* Dialog Title *********************/
20.     AlertDialog dialog = dialogContentBuilder.create();
21.     dialog.setTitle("Welcome to Euroup Cup 2012");
22.     dialog.setIcon(R.drawable.ic_menu_largetiles);
23.     /******************* Dialog Bottom *********************/
24.     dialog.setButton(DialogInterface.BUTTON_POSITIVE,"OK",
25.             new android.content.DialogInterface.OnClickListener() {
26.                 public void onClick(DialogInterface dialog,int which) {
27.                     dialog.dismiss();
28.                 }
29.             });
30.     return dialog;
31. }
```

运行欧洲杯信息采集系统 APP 并单击主界面的小组时，弹出如图 8-7 所示的界面。

图 8-7　欧洲杯小组内国家积分榜

模块三　赛事明细列表的实现

【模块描述】

在图 8-7 所示的积分榜对话框中，单击某个参赛队伍的行信息前面的"∧"图标，即可显示这个国家的赛事明细列表；单击已经展开的某个参赛队伍的行信息前面的"∨"图标，可以将赛事明细折叠起来。

本模块将实现如下操作：

1）将赛事明细以列表形式列出；

2）可折叠的积分榜明细列表的实现。

知识点	技能点
➢ ExpendedListView 组件的用法 ➢ Android 数据适配器的类型及用法 ➢ Android 常见的组件用法	➢ 数据适配器的定义及使用 ➢ 基于 ExpendedListView 实现可折叠的列表

任务1　以列表形式呈现赛事明细

【任务描述】

本任务为以列表形式显示所有比赛的赛事明细。

【任务实施】

1）创建单场比赛的列表布局文件 score_players_layout.xml，该文件用于一场比赛的数据呈现，其相应代码如下所示：

```
1.  <?xml version="1.0" encoding="utf-8"?>
2.  <LinearLayout xmlns:android="http://schemas.android.com/apk/res/android"
3.      android:id="@+id/scoreresultpanel"
4.      android:layout_width="match_parent"
5.      android:layout_height="wrap_content"
6.      android:gravity="left|center_vertical"
7.      android:orientation="horizontal"
8.      android:baselineAligned="false"
9.      android:padding="3dp" >
10.     <ImageView
11.         android:id="@+id/playerstatus_home"
12.         android:layout_width="10dp"
13.         android:layout_height="10dp"
14.         android:layout_marginRight="3dp"
15.         android:contentDescription="@null" />
16.     <TextView
17.         android:id="@+id/txtPlayerName_home"
18.         android:layout_width="0dp"
19.         android:layout_height="wrap_content"
20.         android:gravity="left|center_horizontal"
21.         android:textSize="10sp"
22.         android:layout_weight="1" />
23.     <TextView
24.         android:id="@+id/txtTime_home"
25.         android:layout_width="wrap_content"
26.         android:layout_height="wrap_content"
27.         android:gravity="left|center_horizontal"
28.         android:textSize="10sp"
29.         android:layout_marginRight="5dp" />
30.     <ImageView
31.         android:id="@+id/playerstatus_away"
32.         android:layout_width="10dp"
```

```
33.         android:layout_height = "10dp"
34.         android:layout_marginRight = "3dp"
35.         android:contentDescription = "@null"/>
36.     <TextView
37.         android:id = "@+id/txtPlayerName_away"
38.         android:layout_width = "0dp"
39.         android:layout_height = "wrap_content"
40.         android:gravity = "left|center_horizontal"
41.         android:textSize = "10sp"
42.         android:layout_weight = "1" />
43.     <TextView
44.         android:id = "@+id/txtTime_away"
45.         android:layout_width = "wrap_content"
46.         android:layout_height = "wrap_content"
47.         android:gravity = "left|center_horizontal"
48.         android:textSize = "10sp"
49.         android:layout_marginRight = "5dp" />
50. </LinearLayout>
```

2）创建新布局文件 score_result_layout.xml，用于呈现赛事比分和赛事明细。该布局文件从上而下共包含两个视图：TextView 和 ListView。其中，TextView 用于显示赛事比分，ListView 用于显示赛事明细。其相应代码如下所示：

```
1.  <?xml version = "1.0" encoding = "utf-8"?>
2.  <LinearLayout xmlns:android = "http://schemas.android.com/apk/res/android"
3.      android:id = "@+id/scoreresultpanel"
4.      android:layout_width = "match_parent"
5.      android:layout_height = "wrap_content"
6.      android:gravity = "left|center_vertical"
7.      android:orientation = "vertical"
8.      android:padding = "3dp" >
9.      <TextView
10.         android:id = "@+id/txtScoreResult"
11.         android:layout_width = "match_parent"
12.         android:layout_height = "wrap_content"
13.         android:gravity = "center_vertical|center_horizontal" />
14.     <ListView
15.         android:id = "@+id/gvPlayers"
16.         android:layout_width = "match_parent"
17.         android:layout_height = "0dp"
18.         android:layout_weight = "1"
19.         android:scrollbars = "vertical"
20.         android:overScrollMode = "always"
```

```
21.            android:scrollbarStyle = "outsideOverlay"
22.            android:scrollbarAlwaysDrawVerticalTrack = "true" / >
23.    </LinearLayout >
```

其他程序代码的实现请查看随书配资源，应用运行后所得界面如图 8-8 所示。

图 8-8　赛事明细界面

任务 2　以列表形式呈现可折叠的积分榜明细

【任务描述】

在图 8-7 所示的积分榜对话框中，单击某个参赛队伍的行信息前面的"∧"图标，即可显示这个国家的赛事明细列表；单击已经展开的某个参赛队伍的行信息前面的"∨"图标，可以将赛事明细折叠起来。主视图显示 A 组积分榜，子视图显示 A 组赛事明细。

【任务实施】

1) 创建 PlayerHolder，其相应程序代码如下：

```
1.    public class PlayerHolder {
2.        public TextView result;
3.        public ListView players;
4.    }
```

2）创建 PlayerEventAdapterRowManager 类，实现 AdapterRowManager 接口，其相应程序代码如下所示：

```
1.   public class PlayerEventAdapterRowManager implements AdapterRowManager {
2.       private LayoutInflater inflater = null;
3.       private LinearLayout row = null;
4.       public PlayerEventAdapterRowManager(LayoutInflater inflater) {
5.           this.inflater = inflater;
6.       }
7.       private LinearLayout getChildRow(View convertView) {
8.           if (convertView == null) {
9.               row = (LinearLayout) inflater.inflate(R.layout.score_result_layout,
10.                      null);
11.          } else {
12.              row = (LinearLayout) convertView;
13.          }
14.          return row;
15.      }
16.      private PlayerHolder getPlayerHolderAndSaveIt(View convertView) {
17.          if (convertView != null)
18.              return (PlayerHolder) row.getTag();
19.          PlayerHolder holder = new PlayerHolder();
20.          row.setTag(holder);
21.          holder.result = (TextView) row.findViewById(R.id.txtScoreResult);
22.          holder.players = (ListView) row.findViewById(R.id.gvPlayers);
23.          return holder;
24.      }
25.      public Object getRow(View convertView) {
26.          return getChildRow(convertView);
27.      }
28.      public Object getHolder(View convertView) {
29.          return getPlayerHolderAndSaveIt(convertView);
30.      }
31.  }
```

3）创建 score_result_layout.xml 布局文件，其相应程序代码如下所示：

```
1.   <?xml version = "1.0" encoding = "utf-8"? >
2.   <LinearLayout xmlns:android = "http://schemas.android.com/apk/res/android"
3.       android:id = "@+id/scoreresultpanel"
4.       android:layout_width = "match_parent"
5.       android:layout_height = "wrap_content"
6.       android:gravity = "left|center_vertical"
```

```
7.        android:orientation = "vertical"
8.        android:padding = "3dp" >
9.        <TextView
10.            android:id = "@ + id/txtScoreResult"
11.            android:layout_width = "match_parent"
12.            android:layout_height = "wrap_content"
13.            android:gravity = "center_vertical|center_horizontal" / >
14.        <ListView
15.            android:id = "@ + id/gvPlayers"
16.            android:layout_width = "match_parent"
17.            android:layout_height = "0dp"
18.            android:layout_weight = "1"
19.            android:scrollbars = "vertical"
20.            android:overScrollMode = "always"
21.            android:scrollbarStyle = "outsideOverlay"
22.            android:scrollbarAlwaysDrawVerticalTrack = "true" / >
23.    </LinearLayout >
```

4) 打开 ScoreboardAdapter 适配器，增加子视图显示小组赛事明细，其增加的相应程序代码如下所示：

```
1.    /**
2.     * 获取子视图对象
3.     */
4.    public View getChildView(int groupPosition, int childPosition,
5.            boolean isLastChild, View convertView, ViewGroup parent) {
6.        String teamName = group.getTeams().get(groupPosition).getName();
7.        List < Game > games = this.group.findGamesByAssignedTeam(teamName);
8.        Game game = games.get(childPosition);
9.        PlayerEventAdapterRowManager manager = new PlayerEventAdapterRowManager(
10.               inflater);
11.       LinearLayout row = (LinearLayout) manager.getRow(convertView);
12.       PlayerHolder holder = (PlayerHolder) manager.getHolder(convertView);
13.       setDataForChildRow(holder, game);
14.       return row;
15.   }
16.   /***************************************************************/
17.   private void setDataForChildRow(PlayerHolder holder, Game game) {
18.       List < Map < String, Object > > parameters = new ArrayList < Map < String, Object > > ();
19.       int size = game.getHome_event().getPlayers().size() > game
20.               .getAway_event().getPlayers().size() ? game.getHome_event()
21.               .getPlayers().size() : game.getAway_event().getPlayers().size();
22.       for (int i = 0; i < size; i ++ ) {
```

```java
23.             Map < String, Object > parameter = new HashMap < String, Object > ();
24.             if ( i >= game.getHome_event().getPlayers().size()) {
25.                 parameter.put("icon_home", null);
26.                 parameter.put("name_home", null);
27.                 parameter.put("time_home", null);
28.             } else {
29.                 parameter.put("icon_home", getImageIdentitifier(game
30.                         .getHome_event().getPlayers().get(i).getEvent()));
31.                 parameter.put("name_home", game.getHome_event().getPlayers()
32.                         .get(i).getName());
33.                 parameter.put("time_home", game.getHome_event().getPlayers()
34.                         .get(i).getTime());
35.             }
36.             if ( i >= game.getAway_event().getPlayers().size()) {
37.                 parameter.put("icon_away", null);
38.                 parameter.put("name_away", null);
39.                 parameter.put("time_away", null);
40.             } else {
41.                 parameter.put("icon_away", getImageIdentitifier(game
42.                         .getAway_event().getPlayers().get(i).getEvent()));
43.                 parameter.put("name_away", game.getAway_event().getPlayers()
44.                         .get(i).getName());
45.                 parameter.put("time_away", game.getAway_event().getPlayers()
46.                         .get(i).getTime());
47.             }
48.             parameters.add(parameter);
49.         }
50.         SimpleAdapter adapter = new SimpleAdapter(context, parameters,
51.                 R.layout.score_players_layout, new String[] { "icon_home",
52.                         "name_home", "time_home", "icon_away", "name_away",
53.                         "time_away" }, new int[] { R.id.playerstatus_home,
54.                         R.id.txtPlayerName_home, R.id.txtTime_home,
55.                         R.id.playerstatus_away, R.id.txtPlayerName_away,
56.                         R.id.txtTime_away });
57.         holder.players.setAdapter(adapter);
58.         holder.result.setText(game.getHome_team() + " " + game.getResult() + " " +
59. game.getAway_team());
60.     }
61.     private void setDataForGroupRow(TeamHolder holder, int groupPosition) {
62.         Team team = group.getTeams().get(groupPosition);
63.         holder.sequence.setText(String.valueOf(groupPosition + 1));
64.         holder.name.setImageResource(getImageIdentitifier(team.getName()));
```

```
65.            holder. W. setText(team. getW());
66.            holder. D. setText(team. getD());
67.            holder. L. setText(team. getL());
68.            holder. F. setText(team. getF());
69.            holder. A. setText(team. getA());
70.            holder. Pts. setText(team. getPts());
71.        }
```

从上面的程序代码可以看出：

① 积分榜功能的显示需要通过 getChildView 方法实现，利用当前适配器中的 Group 获取 Team：

  a. 通过 groupPosition 参数了解当前正在处理的是可折叠列表的第几行；

  b. 获取 Group 中的 Teams 集合；

  c. 根据 groupPosition 属性获取一个 Team 对象。

② 利用 Team 中的 findGamesByAssignedTeam 方法获取与该球队有关的比赛集合。

③ 根据 Android 处理可折叠列表子项的顺序和原理获取想用的 Game 对象：

  a. 通过 childPosition 参数了解当前正在处理的是可折叠列表子项的第几行；

  b. 根据 childPosition 属性获取一个 Game 对象。

④ 将 Game 中的 home_name、resutl、away_name 显示在比分栏上。

运行应用后得到的效果图如图 8-9 所示。

图 8-9 可折叠的积分榜明细列表界面

项目九 新闻客户端 APP 应用

本项目属于综合项目，除了用到第一篇介绍的常用 Android 技术知识点外，还涉及 Android 网络编程、Android 简单动画、JSON 轻量级数据解析等技术。本项目主要实现新闻客户端 APP 应用，需要访问服务器端应用程序"新闻发布系统"。本书只给出服务器端应用程序的部署步骤与方法，不涉及服务器端应用程序的开发（具体源码可在本书配套资源，即项目九文件夹中服务器端文件夹里找到）。在开发新闻客户端 APP 时，只需要访问服务器端给出的接口即可实现 APP 应用功能。本项目分为 3 个模块：

1）用户登录功能的实现；
2）新闻栏目列表功能的实现；
3）新闻栏目列表功能的实现。

【知识目标】

➢ Android 组件 Activity 的灵活运用
➢ 了解 Android 动画，能进行简单的 Android 动画编程
➢ 自定义数据适配器，结合 ListView 组件灵活使用适配器，将数据显示到界面列表项
➢ 掌握 Android 网络编程的方法和步骤
➢ 掌握基于 HTTP 与服务器建立连接的方法
➢ 掌握 JSON 轻量级数据解析技术的相关知识点
➢ 掌握基于 JSON 技术实现新闻栏目及新闻数据的封装和解析方法
➢ 掌握 Android 网络资源（图片资源）的访问方法

【模块分解】

➢ 用户登录功能的实现
➢ 新闻栏目列表功能的实现

模块一 用户登录功能的实现

【模块描述】

新闻客户端 APP 应用分为用户登录、新闻栏目及新闻 3 个子项目。本模块的用户登录子项目，主要基于线性布局 LinearLayout、主题 Theme、ImageView 组件、Android 动画、网络资源访问、JSON 数据封装技术实现用户登录功能。

本模块将实现如下操作：
1）准备工作，即对服务器端应用程序的部署；

2）用户登录功能主界面的实现；

3）基于 HTTP 访问服务器实现 APP 客户端用户登录功能。

知识点	技能点
➢ 服务器端 Web 容器的安装步骤及服务启动 ➢ 服务器端应用程序的部署步骤及方法 ➢ Android 简单动画设置 ➢ Activity 组件的灵活运用 ➢ Android 基本 UI 组件运用 ➢ Android 的事件处理机制 ➢ Android UI 线程 ➢ Android 网络资源访问	➢ Web 容器的安装及服务启动步骤 ➢ 服务器端应用程序的部署方法 ➢ 网络访问工具及配置文件创建的必要性和创建方法 ➢ 基于 HTTP 访问服务器 ➢ JSON 数据解析技术的简单运用

任务 1　准备工作：服务器端应用程序的部署

【任务描述】

新闻客户端 APP 应用需要访问服务器端应用程序，因此在实现 APP 应用之前需要搭建服务器端运行环境及部署应用程序。首先需要安装 Web 容器，其次需要将服务器端应用程序部署到容器中并启动服务。本应用程序采用 Tomcat 6.0 应用服务器。

【任务实施】

1）安装 Tomcat 6.0 应用服务器。

① 下载 Tomcat 软件。Tomcat 是免费的开源软件，进入如下地址下载：http://tomcat.apache.org/，本书使用 Tomcat 6.0 版本。在安装之前要确保 JDK 成功安装。

② 下载后解压，双击图标运行文件，即可进入欢迎安装界面，如图 9-1 所示。

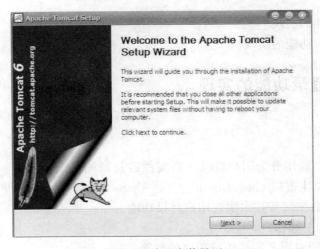

图 9-1　欢迎安装界面

③ 单击上图中的 Next 按钮，进入如图 9-2 所示的协议授权界面。

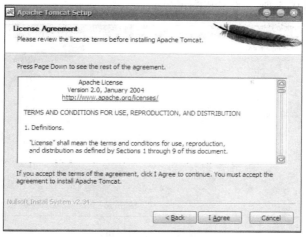

图 9-2　协议授权界面

④ 单击上图中的 I Agree 按钮，进入安装组件选择界面，如图 9-3 所示。

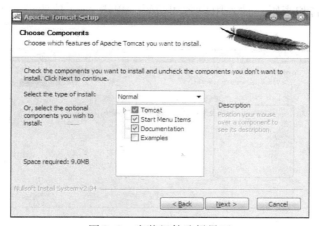

图 9-3　安装组件选择界面

⑤ 直接单击上图中的 Next 按钮，进入路径选择界面，如图 9-4 所示。

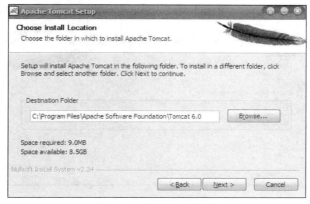

图 9-4　路径选择界面

241

⑥ 单击 Browse 按钮，修改安装路径（将本任务安装在"C:\Tomcat 6.0"路径下），单击 Next 按钮进入如图 9-5 所示的安装配置界面。

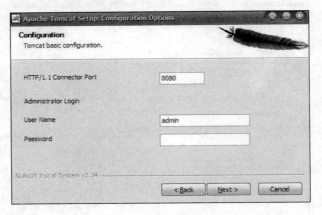

图 9-5　安装配置界面

⑦ 上图中"8080"为 Tomcat 的端口号，这里采用默认设置。User Name 及 Password 是管理 Tomcat 的用户名和密码。输入用户名和密码，单击 Next 按钮进入如图 9-6 所示的 JRE 配置界面。

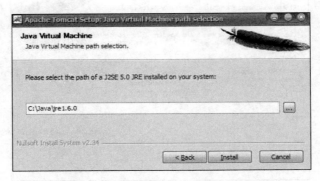

图 9-6　JRE 配置界面

⑧ 在上图中首先选择 JRE 的安装路径（注：如果在安装 Tomcat 之前 JDK 已成功安装，则会自动找到 JRE 的安装路径），单击 Install 按钮进入如图 9-7 所示的界面。

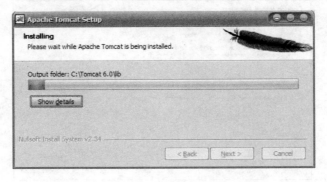

图 9-7　安装进度界面

安装完成后，如果在浏览器地址栏中输入"http://localhost:8080"，能看到如图9-8所示的界面，则Tomcat安装成功。

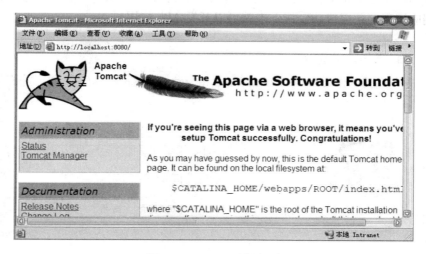

图9-8　Tomcat运行主页

2）部署应用程序到Tomcat 6.0应用服务器的webapps下。

① 将随书配资源中对应内容中的"服务器端"文件夹打开，复制"News_PC"项目代码到安装好的Tomcat目录"webapps"下（本任务安装路径为C:\Tomcat 6.0\webapps）。

② 安装SQLServer 2005数据库（步骤省略），将随书配资源中对应内容中的"db"文件夹打开，对"NewsSystem"数据库进行附加操作，将访问数据库的用户名和密码都设置为"sa"，启用数据库"TCP/IP"协议，重新启动数据库服务。

③ 启动Tomcat服务，单击Start按钮即可，如图9-9所示。

图9-9　启动Tomcat

④ 打开 IE 浏览器，在地址栏中输入"http://localhost:8080/News_PC/jsp/login.jsp"，在如图 9-10 所示的界面中输入用户名和密码分别为"shl"和"12345"。如果登录成功，则服务器端应用程序部署成功，接下来就可以开发 APP 应用了。

图 9-10　服务器端运行界面

任务 2　编写配置文件及网络访问工具

【任务描述】

APP 应用需要经常访问网络资源，为了实现代码重用，需要编写一个将 Inputstream 流内容转换成 byte[] 的工具类。另外从 APP 部署的便利性考虑，编写一个配置文件，用于封装 IP 地址及网络资源路径。

1）创建一个新的 Android 应用程序。
2）编写系统配置文件。
3）编写 Inputstream 流转换工具类。

【任务实施】

1）创建 Android 应用程序"News_App"，选择 Android 4.2 版本，在 src 文件夹下创建包"czmec.cn.shl.news"，在这个包下分别创建"activity""entity""config"和"util"4 个包，其中，后两个包用来管理系统配置文件和工具，如图 9-11 所示。

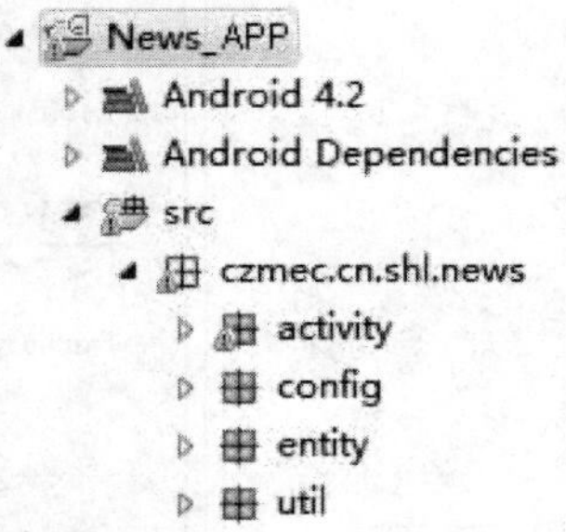

2）在 config 包下创建"Config.java"文件，其相应程序代码为：

图 9-11　"News_APP"应用结构

1. /*
2. 　* 配置类

244

```
3.        * PC 服务器的 IP 地址等
4.        */
5.   public class Config {
6.       public static String IPAddress = "172.20.250.210:8080/News_PC";
7.       public static String ServerAddress = "http://" + IPAddress + "/servlet/";
8.       public static String ImageAddress = "http://" + IPAddress + "/photos/pic/";
9.   }
```

其中，IPAddress 中的"172.20.250.210"因此部署应用程序的服务器端 IP 地址，"8080"为 Tomcat 的通信端口号，ServerAddress 为 APP 应用访问的资源地址，ImageAddress 为 APP 应用访问的图片资源地址。

3）编写 Inputstream 流转换工具类，此工具类将在用户登录功能实现中使用。在 util 包下创建"SynsHttp.java"文件，其相应程序代码为：

```
1.   public class SynsHttp {
2.       public static byte[] readParse(String urlPath) throws Exception {
3.           ByteArrayOutputStream outStream = new ByteArrayOutputStream();
4.           byte[] data = new byte[1024];
5.           int len = 0;
6.           //根据 URL 获取网络资源
7.           //通过 URL 进行地址解析
8.           URL url = new URL(urlPath);
9.           //打开网络链接
10.          HttpURLConnection conn;
11.          conn = (HttpURLConnection) url.openConnection();
12.          //设置请求的方式
13.          conn.setRequestMethod("GET");
14.          //设置请求时间的上限
15.          conn.setConnectTimeout(5 * 1000);
16.          //获取网络中的数据流
17.          InputStream is = conn.getInputStream();
18.          //将 Inputstream 流转换成二进制字节码
19.          InputStream inStream = conn.getInputStream();
20.          while ((len = inStream.read(data)) != -1) {
21.              outStream.write(data, 0, len);
22.          }
23.          inStream.close();
24.          return outStream.toByteArray();
25.      }
26.  }
```

上述程序代码定义了一个 readParse 方法，此方法的返回值为 byte[]，传进来的参数为网络请求路径，第 8 行代码定义了一个 URL 对象，第 11 行代码通过调用 url 的 openConnection 方法获取网络链接对象 HttpURLConnection，设置网络访问请求方式为"GET"方式，设置网络访问时限为 5 s，然后调用 HttpURLConnection 的 getInputStream 方法获取 InputStream 输入流。

任务3　构建用户登录界面

【任务描述】

本任务是基于 Android 基础 UI 实现用户登录界面。

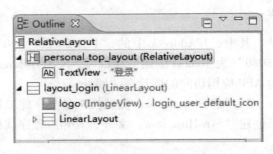

图 9-12　登录界面布局

【任务实施】

在"acticity"包下创建名称为"LoginActivity"的用户登录类，采用相对布局管理器实现，具体界面布局如图 9-12 所示。

登录界面对应的"activity_login.xml"文件，其相应程序如下所示：

```
1.  <?xml version = "1.0" encoding = "utf-8"?>
2.  <RelativeLayout xmlns:android = "http://schemas.android.com/apk/res/android"
3.      android:layout_width = "match_parent"
4.      android:layout_height = "match_parent"
5.      android:background = "@drawable/android_layout_bg" >
6.      <RelativeLayout
7.          android:id = "@+id/personal_top_layout"
8.          android:layout_width = "match_parent"
9.          android:layout_height = "wrap_content"
10.         android:layout_alignParentTop = "true"
11.         android:background = "@drawable/android_title_bg" >
12.         <TextView
13.             android:layout_width = "wrap_content"
14.             android:layout_height = "wrap_content"
15.             android:layout_alignParentLeft = "true"
16.             android:layout_centerVertical = "true"
17.             android:layout_margin = "10dp"
18.             android:text = "@string/login"
19.             android:textColor = "@color/white"
20.             android:textSize = "@dimen/medium_text_size" />
21.     </RelativeLayout>
22.     <LinearLayout
23.         android:id = "@+id/layout_login"
24.         android:layout_width = "fill_parent"
25.         android:layout_height = "fill_parent"
```

```
26.            android:layout_marginTop = "30dp"
27.            android:orientation = "vertical"   >
28.         < ImageView
29.              android:id = "@ + id/logo"
30.              android:layout_width = "90dp"
31.              android:layout_height = "90dp"
32.              android:layout_gravity = "center_horizontal"
33.              android:layout_marginTop = "20dp"
34.              android:background = "@drawable/login_user_icon_bg"
35.              android:src = "@drawable/login_user_default_icon" / >
36.         < LinearLayout
37.              android:layout_width = "match_parent"
38.              android:layout_height = "wrap_content"
39.              android:background = "@drawable/android_layout_bg"
40.              android:orientation = "vertical">
41.              < LinearLayout style = "@style/PersonalMainLayoutStyle"  >
42.                   < LinearLayout
43.                        android:layout_width = "fill_parent"
44.                        android:layout_height = "wrap_content"
45.                        android:layout_margin = "5dp"
46. android:background = "@drawable/more_activity_item_selector_bottom_corners"
47.                        android:orientation = "horizontal"
48.                        android:padding = "10sp"   >
49.                        < TextView
50.                             android:id = "@ + id/tv_login_name_title"
51.                             android:layout_width = "wrap_content"
52.                             android:layout_height = "wrap_content"
53.                             android:text = "@string/loginacount"
54.                             android:textColor = "@android:color/black"
55.                             android:textSize = "18.0sp" / >
56.                        < EditText
57.                             android:id = "@ + id/login_name"
58.                             android:layout_width = "163dp"
59.                             android:layout_height = "wrap_content"
60.                             android:background = "@null"
61.                             android:ems = "10"
62.                             android:focusableInTouchMode = "true"
63.                             android:padding = "5.0dip"
64.                             android:paddingRight = "30dp"
65.                             android:textColor = "#ff3b3b3b"
66.                             android:textSize = "16.0sp"   >
67.                        </EditText >
```

```
68.                </LinearLayout>
69.                <View style = "@style/PersonalLine" />
70.                <LinearLayout
71.                    android:layout_width = "fill_parent"
72.                    android:layout_height = "wrap_content"
73.                    android:layout_margin = "5dp"
74.    android:background = "@drawable/more_activity_item_selector_bottom_corners"
75.                    android:orientation = "horizontal"
76.                    android:padding = "10sp"  >
77.                    <TextView
78.                        android:id = "@+id/tv_loginpassword"
79.                        android:layout_width = "wrap_content"
80.                        android:layout_height = "wrap_content"
81.                        android:text = "@string/loginpassword"
82.                        android:textColor = "@android:color/black"
83.                        android:textSize = "18.0sp" />
84.                    <EditText
85.                        android:id = "@+id/login_password"
86.                        android:layout_width = "163dp"
87.                        android:layout_height = "wrap_content"
88.                        android:background = "@null"
89.                        android:ems = "10"
90.                        android:focusableInTouchMode = "true"
91.                        android:inputType = "textPassword"
92.                        android:padding = "5.0dip"
93.                        android:paddingRight = "30dp"
94.                        android:textColor = "#ff3b3b3b"
95.                        android:textSize = "16.0sp"  >
96.                    </EditText>
97.                </LinearLayout>
98.            </LinearLayout>
99.            <Button
100.               android:id = "@+id/login_btn"
101.               android:layout_width = "fill_parent"
102.               android:layout_height = "wrap_content"
103.               android:layout_marginBottom = "14dp"
104.               android:layout_marginLeft = "15dp"
105.               android:layout_marginRight = "15dp"
106.               android:layout_marginTop = "10dp"
107.               android:background = "@drawable/android_title_bg"
108.               android:gravity = "center"
109.               android:text = "@string/login_btn"
```

```
110.                    android:textColor = "#fff"
111.                    android:textSize = "20sp"
112.                    android:onClick = "loginoperate" / >
113.            < RelativeLayout
114.                    android:id = " @ + id/layout_login_userinfo"
115.                    android:layout_width = "fill_parent"
116.                    android:layout_height = "wrap_content"
117.                    android:layout_marginTop = "10dp"
118.                    android:orientation = "horizontal"   >
119.            < /RelativeLayout >
120.            < Button
121.                    android:id = " @ + id/register_btn"
122.                    android:layout_width = "fill_parent"
123.                    android:layout_height = "40sp"
124.                    android:layout_marginLeft = "15dp"
125.                    android:layout_marginRight = "15dp"
126.                    android:layout_marginTop = "5dp"
127.                    android:background = "@ drawable/login_register_bg"
128.                    android:gravity = "center"
129.                    android:text = "@ string/registerfree"
130.                    android:textSize = "20sp" / >
131.        < /LinearLayout >
132.    < /LinearLayout >
133. < /RelativeLayout >
```

任务4　基于 HttpURLConnection 实现 APP 客户端用户登录

【任务描述】

实现 APP 客户端用户登录功能。

【任务实施】

打开"activity"包下的"LoginActivity.java",其相应程序代码如下：

```
1.  //定义 APP 页面中的用户名、密码以及登录按钮、注册按钮
2.      private EditText t_login_name,t_login_password;
3.      private Button login_btn,register_btn;
4.      private int uid;
5.      private String uName ,uPass,path,result;
6.  //定义初始化的方法
7.      private void initData( )
8.      {
9.          t_login_name = (EditText) this.findViewById( R. id. login_name);
```

```
10.            t_login_password = (EditText) this.findViewById(R.id.login_password);
11.            login_btn = (Button) this.findViewById(R.id.login_btn);
12.            register_btn = (Button) this.findViewById(R.id.register_btn);
13.        }
14.        public void loginoperate(View v)
15.        {
16.            uName = t_login_name.getText().toString();//用户输入的用户名
17.            uPass = t_login_password.getText().toString();//用户输入的密码
18.            if(uName.trim().length() == 0 || uPass.trim().length() == 0)
19.            {
20.                //提示用户名和密码不能为空
21.                Toast.makeText(LoginActivity.this, R.string.no_null, Toast.LENGTH_SHORT).show();
22.            }
23.            else
24.            {
25.                try {
26.                    uName = URLEncoder.encode(uName, "gbk");//将用户名进行编码转换
27.                    uPass = URLEncoder.encode(uPass, "gbk");
28.                    path = Config.ServerAddress + "UserLoginAppServlet?userName=" + uName
29.                        + "&userPassword=" + uPass;
30.                    //开启一个新的线程
31.                    new Thread(new Runnable()
32.                    {
33.                        @Override
34.                        public void run() {
35.                            try {
36.                                //将 Inputstream 流转换成二进制字节码
37.                                byte[] data = StreamTools.SynsHttp(path);
38.                                //将这个字节转换成 string
39.                                String jsonstring = new String(data);
40.                                //将 jsonstring 转换成 JSON 对象
41.                                JSONObject jo = new JSONObject(jsonstring);
42.                                result = jo.getString("msg");
43.                                uid = jo.getInt("uid");
44.                                //开启 UI 线程
45.                                runOnUiThread(new Runnable()
46.                                {
47.                                    @Override
48.                                    public void run() {
49.                                        Toast.makeText(LoginActivity.this, result, Toast.LENGTH_SHORT).show();
50.                                        if(result.equals("userLoginSuccess"))
```

250

```
51.                            }
52.                    Intent intent = new Intent(LoginActivity.this,MainActivity.class);
53.                                    startActivity(intent);
54.                                    //增加动画效果
55.         LoginActivity.this.overridePendingTransition(R.anim.in_translate_left_one,
56.   R.anim.in_translate_left_two);
57.                            }
58.                    }
59.                    });
60.            } catch (Exception e) {
61.                    //TODO Auto-generated catch block
62.                    e.printStackTrace();
63.            }
64.            }
65.        }
66.                    ).start();
67.    } catch (UnsupportedEncodingException e) {
68.            //TODO Auto-generated catch block
69.            e.printStackTrace();
70.    }
71.   }
72.  }
73. }
```

上述代码的第28、29行，即 path = Config.ServerAddress + "UserLoginAppServlet?userName = " + uName + "&userPassword = " + uPass 是服务器端用户登录访问 API 的接口，当程序运行时上述代码中的 path 将被赋值：http://192.168.1.106:8080/News_NewBee_PC/servlet/UserLoginAppServlet? userName = zel&userpass = 1111。其中，路径前缀"http"为超文本传输协议，"192.168.1.106"为服务器 IP 地址，"News_NewBee_PC/servlet/UserLoginAppServlet"为请求的资源，"?userName = zel&userpass = 1111"为 APP 客户端用户输入的登录名和密码，分别为"zel"和"1111"。

上述代码的第53行，即 Intent intent = new Intent(LoginActivity.this, MainActivity.class) 中的"MainActivity"为 APP 应用登录成功后的系统主界面，其对应的 XML 布局文件的文件名为"activity_main.xml"，其相应程序代码如下所示：

```
1. <RelativeLayout xmlns:android = "http://schemas.android.com/apk/res/android"
2.     xmlns:tools = "http://schemas.android.com/tools"
3.     android:layout_width = "match_parent"
4.     android:layout_height = "match_parent"
5.     android:background = "#FFFFFF"
6.     android:orientation = "vertical" >
7.     <Button
```

```
8.          android:id = "@+id/news_ch_all"
9.          android:layout_width = "match_parent"
10.         android:layout_height = "98dp"
11.         android:layout_above = "@+id/news_ch"
12.         android:layout_alignParentLeft = "true"
13.         android:layout_marginBottom = "52dp"
14.         android:background = "@drawable/index_bn_all"
15.         android:onClick = "getInAll" />
16.     <Button
17.         android:id = "@+id/news_ch"
18.         android:layout_width = "match_parent"
19.         android:layout_height = "88dp"
20.         android:layout_alignParentBottom = "true"
21.         android:layout_alignParentLeft = "true"
22.         android:layout_marginBottom = "138dp"
23.         android:background = "@drawable/index_bn_category"
24.         android:onClick = "getIn" />
25. </RelativeLayout>
```

上述程序代码第24行中黑体字"getIn"是定义的按钮单击事件响应方法，具体可以参见本项目模块二任务了"JSON 轻量级数据解析技术实现新闻栏目列表功能"。

上述程序代码的第55、56行，即 LoginActivity. this. overridePendingTransition(R. anim. in_translate_left_one , R. anim. in_translate_left_two)，是为加载 Android 位置平移的动画效果。其中，anim. in_translate_left_one 和 anim. in_translate_left_two 是在 res 资源包下的"anim"文件夹（anim 文件夹可以自己创建）中定义的动画资源。

anim. in_translate_left_one. xml 文件其相应程序代码如下所示：

```
1.  <translate
2.      android:fromXDelta = "100%"
3.      android:toXDelta = "0"
4.      android:fromYDelta = "0"
5.      android:toYDelta = "0"
6.      android:duration = "2000"
7.  xmlns:android = "http://schemas.android.com/apk/res/android" >
8.      public static String ImageAddress = "http://" + IPAddress + "/photos/pic/";
9.  }
```

anim. in_translate_left_two. xml 文件其相应程序代码如下所示：

```
1.  <translate
2.      android:fromXDelta = "0"
3.      android:toXDelta = " -100%"
4.      android:fromYDelta = "0"
```

5.　　　　android:toYDelta = "0"
6.　　　　android:duration = "2000"
7.　　xmlns:android = "http://schemas.android.com/apk/res/android" >
8.　　public static String ImageAddress = "http://" + IPAddress + "/photos/pic/";
9.　}

登录界面的运行如图9–13所示。

当用户名或密码没有输入时，将弹出Toast提示"用户名或密码不能为空"，当登录成功时界面将跳转到"MainActivity"界面。

【知识学习】 Android 平移动画

Android 平台提供了一套完整的动画框架。在 Android 3.0 之前有两种动画，一种是补间动画（Tween Animation），另一种叫逐帧动画（Frame Animation），也称 Drawable Animation。Android 3.0 以后增加了属性动画 Property Animation。上述用户登录模块使用了 Animation 的 translate 位置平移动画，其使用时非常简单，相关属性如表9–1所列。

图9–13　用户登录界面

表9–1　translate 位置平移动画效果

序　号	属　性	含　义
1	fromXDelta	属性为动画起始时 X 坐标上的位置
2	toXDelta	属性为动画结束时 X 坐标上的位置
3	fromYDelta	属性为动画起始时 Y 坐标上的位置
4	toYDelta	属性为动画结束时 Y 坐标上的位置
5	duration	属性为动画持续时间，时间以毫秒为单位

注意：

1) 没有指定 fromXType toXType、fromYType toYType 时候，默认是以自己为相对参照物。

2) 在 anim 目录中新建的动画资源文件名要小写，如 my_anim.xml。

模块二　新闻栏目列表功能的实现

【模块描述】

本模块为新闻栏目子项目，主要基于线性布局管理器、数据适配器、ListView 组件、Android 网络资源访问、JSON 数据封装技术实现新闻栏目列表功能。

本模块将实现如下操作。

1) 准备工作：测试服务器端新闻栏目列表 API 接口。

2) 构建新闻栏目列表界面，考虑到界面编码的可重用性，将新闻栏目列表界面的"抬头标题"定义为一个独立的界面文件。

3）定义"抬头标题"独立界面文件 title.xml。
4）创建 JSON 数据解析工具类 ParseJson。
5）创建新闻栏目列表的数据适配器 NewsBarlistAdapter。
6）实现新闻栏目列表功能。

知识点	技能点
➢ 基于数据适配器填充 ListView 的方法 ➢ 功能模块化及代码重用设计思想 ➢ 自定义数据适配器 ➢ ListView 组件中列表样式的灵活定义 ➢ Android 基本 UI 组件运用 ➢ ListView 组件的 OnItemClickListener 事件处理机制 ➢ Android UI 中 include 的运用 ➢ Android 网络资源访问	➢ "抬头标题"独立界面文件 title.xml 的定义及引用 ➢ 数据解析工具类 ParseJson 的定义 ➢ 数据适配器 NewsBarlistAdapter 的定义及 ListView 组件数据填充 ➢ 新闻栏目列表界面的定义

任务1 准备工作：测试服务器端新闻栏目列表 API 接口

【任务描述】

新闻客户端 APP 应用之新闻栏目列表的实现需要访问服务器端应用程序相关 API 接口，因此在实现新闻栏目列表功能之前，首先要确认服务器端应用程序是否部署成功并检验 API 接口是否可用。

【任务实施】

1）按照模块1的任务1中的步骤确认服务器端 Tomcat 6.0 应用服务器成功启动。
2）服务器端新闻栏目列表 API 访问接口的地址如下：

http://192.168.1.106:8080/News_PC/servlet/GetAllNewsBarAppServlet

在浏览器地址栏中输入上述地址后按〈Enter〉键，出现如图 9-14 所示的界面，则 API 接口可用。

图 9-14 新闻栏目访问接口 API 的运行界面

上图中的数据来自服务器中的数据库，数据格式即为 JSON 数据格式。可以看出，上述所有新闻栏目数据被封装在名称为"newsbarlist"的 JSON 对象中。这个 JSON 对象是数组格式，里面封装了 7 个新闻栏目条目，每个新闻栏目条目包含 7 对键/值对，如第一个条目中的 key 键"creatDate"对应的 value 值为"2013 - 6 - 23"。在 APP 应用的新闻栏目列表功能中将编程实现对上述 JSON 数据进行解析。

任务 2　构建新闻栏目列表界面

【任务描述】

新闻栏目列表界面是将新闻栏目以列表的形式展现的界面。为了实现界面代码的可重用性，将本界面中的"抬头标题"设计为一个独立的布局文件，这样就可以在新闻栏目列表和新闻列表界面中引用，代码更加简洁。另外，为了美化及定义新闻栏目列表项的外观和布局，需要创建一个列表项文件。本任务可以拆分为 3 个操作：

1）创建"抬头标题"独立布局文件 title.xml；
2）创建列表项文件 newsbar_list_item.xml；
3）在新闻栏目列表界面中引用 title.xml。

【任务实施】

1）在 res 文件夹的"layout"中创建"抬头标题"独立布局文件 title.xml，页面代码布局如图 9-15 所示。

可以看出，独立布局文件中包含一个用于显示"标题"的 TextView 组件和返回上一界面的 Button 按钮。定义按钮属性"android:onClick = "back""和"back"单击事件方法将在本模块的任务 3 中实现。

图 9-15　独立布局文件 title.xml

2）在 res 文件夹的"layout"中创建列表项文件 newsbar_list_item.xml，其相应程序代码如下：

```
1.    < RelativeLayout xmlns:android = "http://schemas.android.com/apk/res/android"
2.        android:layout_width = "fill_parent"
3.        android:layout_height = "fill_parent" >
4.            < ImageView
5.            android:id = "@ + id/x1"
6.            android:layout_width = "wrap_content"
7.            android:layout_height = "wrap_content"
8.            android:layout_alignParentRight = "true"
9.            android:layout_alignParentTop = "true"
10.           android:src = "@ drawable/product_in" / >
11.       < TextView
12.           android:id = "@ + id/newsbar_item_title"
13.           android:layout_width = "wrap_content"
14.           android:layout_height = "wrap_content"
```

```
15.           android:layout_marginTop = "20dp"
16.           android:layout_alignParentLeft = "true"
17.           android:layout_marginLeft = "30dp"
18.           android:text = "TextView"
19.           android:textColor = "#84C1FF"
20.           android:textSize = "15sp" />
21.  </RelativeLayout>
```

可见,列表中的每个列表项左边为一个 TextView 组件,用于显示每个新闻栏目名称;右边为一个 ImageView 组件,用于显示">"图标。列表项文件 newsbar_list_item.xml 将在新闻栏目数据适配器中被引用。

3) 创建名为"AllNewsBarActivity"的 Activity 类,对应的布局文件为"activity_all_newsbar.xml",页面代码布局如图 9-16 所示。

在线性布局管理器中有 include 和 ListView 两个组件,使用 include 引用 title.xml 文件,即 <include layout = "@layout/**title**" … >。

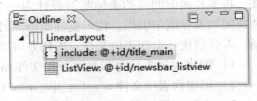

图 9-16 新闻栏目布局文件

任务 3 用 JSON 轻量级数据解析技术实现新闻栏目列表功能

【任务描述】

新闻栏目列表功能的实现相对较复杂,不仅要在 APP 应用的客户端获取服务器端的新闻栏目列表数据,还要将获取的数据在列表中显示,具体可以细分为以下 3 个操作:

1) 创建数据解析工具类 ParseJson,解析从服务器获取的新闻栏目列表数据;
2) 创建新闻栏目列表的数据适配器 NewsBarlistAdapter;
3) 实现新闻栏目列表功能。

【任务实施】

1) 在"czmec.cn.shl.news.entity"包下创建 NewBar 类,用来封装新闻栏目列表数据。
2) 在"czmec.cn.shl.news"包下创建 engine 包,基于 engine 包创建数据引擎工具类 ParseJson,解析从服务器获取的新闻栏目列表数据,其相应代码如下:

```
1.  public class ParseJson {
2.  //解析新闻栏目
3.      public static List < NewsBar > getNewsBarList( String str ) {
4.          List < NewsBar > list = new ArrayList < NewsBar > ( );
5.          try {
6.              JSONObject jsonobject = new JSONObject( str );
7.              JSONArray dislist = jsonobject.getJSONArray( "newsbarlist" );
8.              NewsBar newsbar;
9.              for ( int i = 0; i < dislist.length( ); i + + ) {
10.                 JSONObject jo = ( JSONObject ) dislist.opt( i );
```

```
11.                    newsbar = new NewsBar( );
12.                    newsbar. setNewsTypeID( jo. getString( "newsTypeID" ) );
13.                    try {
14.                        newsbar. setTitleName( jo. getString( "titleName" ) );
15.                    } catch ( Exception e ) {
16.                        //TODO Auto – generated catch block
17.                        e. printStackTrace( );
18.                    }
19.                    list. add( newsbar);
20.                }
21.            } catch ( JSONException e ) {
22.                e. printStackTrace( );
23.                return null;
24.            }
25.            return list;
26.        }
27.    }
```

上述程序代码的第 6 行，即 JSONObject jsonobject = new JSONObject(str) 创建了一个 JSON 对象 jsonobject，第 7 行调用 jsonobject 的 getJSONArray 方法获取 JSON 数组对象 "newsbarlist"，"newsbarlist" 必须和图 9-14 中的 "newsbarlist" 相同。第 9 行代码实现了对 JSON 数组对象 newsbarlist 的遍历操作；第 10 行调用其 opt 方法获取每个 JSON 对象 jo；第 12 行使用 jo. getString("newsTypeID") 获取每个对象中的属性值，其中 "newsTypeID" 和图 9-14 中的 "newsTypeID" 键对应。

3）在 "czmec. cn. shl. news" 包下创建 adapter 包，基于 adapter 包创建新闻栏目列表的数据适配器 NewsBarlistAdapter 类，其相应代码如下：

```
1.  public class NewsBarlistAdapter extends BaseAdapter {
2.      private List < NewsBar > newsbarList;
3.      private LayoutInflater inflater;
4.      public NewsBarlistAdapter( Context context , List < NewsBar > newsbarList)
5.      {
6.          inflater = LayoutInflater. from( context);
7.          this. newsbarList = newsbarList;
8.      }
9.      public int getCount( ) {
10.         return newsbarList. size( );
11.     }
12.     public Object getItem( int position) {
13.         return position;
14.     }
15.     public long getItemId( int position) {
```

```
16.            return position;
17.        }
18.        public View getView(int position, View convertView, ViewGroup parent) {
19.            if (convertView == null)
20.            {
21.                convertView = inflater.inflate(R.layout.newsbar_list_item, null);
22.            }
23.            TextView tv = (TextView) convertView.findViewById(R.id.newsbar_item_title);
24.            tv.setText(newsbarList.get(position).getTitleName());
25.            return convertView;
26.        }
27.    }
```

上述代码的第 21 行是通过 inflater 的 inflate 加载列表项布局文件 newsbar_list_item.xml，第 23 行代码通过调用 findViewByld 方法获取 id 为 "newsbar_item_title" 的 TextView，此组件用于显示新闻栏目名称。

4）打开 "MainActivity.java" 类，添加 "分类新闻" 按钮单击事件的方法 "getIn"，将当前界面跳转到新闻栏目列表界面 "AllNewsBarActivity"，其相应程序代码如下：

```
1.  public void getIn(View v)
2.  {
3.      Intent intent = new Intent(MainActivity.this, AllNewsBarActivity.class);
4.      startActivity(intent);
5.      //增加动画效果
6.      MainActivity.this.overridePendingTransition(R.anim.in_translate_left_one, R.anim.in_translate_left_two);
7.  }
```

5）打开 "AllNewsBarActivity.java" 类，实现新闻栏目列表功能，其相应程序代码如下。

```
1.  public class AllNewsBarActivity extends Activity {
2.      private ListView listview;
3.      private List<NewsBar> newsBarList;
4.      private NewsBarlistAdapter adapter;
5.      protected void onCreate(Bundle savedInstanceState) {
6.          super.onCreate(savedInstanceState);
7.          setContentView(R.layout.activity_all_newsbar);
8.          listview = (ListView) this.findViewById(R.id.newsbar_listview);
9.          TextView tv = (TextView) this.findViewById(R.id.title_name);
10.         tv.setText("新闻分类");
11.         new Thread(new Runnable() {
12.             public void run() {
13.                 getNewsAllBar();
```

```
14.            }
15.        }).start();
16.    }
17.    public void back(View v)
18.    {
19.        Intent intent = new Intent(AllNewsBarActivity.this,MainActivity.class);
20.        startActivity(intent);
21. overridePendingTransition(R.anim.back_translate_right_one,R.anim.back_translate_right_two);
22.    }
23.    public void getNewsAllBar()
24.    {
25.        String url = Config.ServerAddress + "GetAllNewsBarAppServlet";
26.        byte[] data;
27.        try{
28.            data = SynsHttp.readParse(url);
29.            String jsonStr = new String(data);
30.            newsBarList = ParseJson.getNewsBarList(jsonStr);
31.            for(NewsBar newsbar:newsBarList)
32.            {
33.                System.out.println("打印:" + newsbar.getTitleName());
34.            }
35.        }catch(Exception e){        e.printStackTrace();
36.        }
37.        runOnUiThread(new Runnable(){
38.            public void run(){
39.                //创建一个自定义的数据适配器
40.                adapter = new NewsBarlistAdapter(AllNewsBarActivity.this,newsBarList);
41.                listview.setAdapter(adapter);
42.                listview.setOnItemClickListener(new OnItemClickListener(){
43.                    public void onItemClick(AdapterView<?> arg0,View arg1,int arg2,long arg3){
44.                        System.out.println("-----被单击的Item是---->>" + arg2);
45.                        Bundle bu = new Bundle();
46.                        bu.putString("newsBarID",newsBarList.get(arg2).getNewsTypeID());
47.                        Toast.makeText(AllNewsBarActivity.this,"你单击的栏目是:
48.                        id:" + newsBarList.get(arg2).getNewsTypeID() + ",栏目名称是为"
49.                        + newsBarList.get(arg2).getTitleName(),4000).show();
50.                        bu.putString("NewsBarName",newsBarList.get(arg2).getTitleName());
51.                        Intent intent = new Intent(AllNewsBarActivity.this,AllNewsBarActivity.class);
52.                        intent.putExtras(bu);
53.                        startActivity(intent);
54.                        //增加动画效果
55.                        AllNewsBarActivity.this.overridePendingTransition(R.anim.in_translate_left_one,
```

```
56.                R. anim. in_translate_left_two);
57.                    }
58.                });
59.                }
60.            });
61.        }
62. }
```

运行 APP，应用登录成功后单击"分类新闻"图片按钮，将进入新闻栏目列表界面，如图 9-17 所示。

图 9-17 新闻栏目列表运行界面

单击上图中"新闻分类"左边的"＜"按钮，可以退回 APP 应用的主界面，界面跳转功能的实现是采用 Android 位置平移动画的方式，需要在 anim 文件夹中定义"back_translate_right_one.xml"和"back_translate_right_two.xml"两个资源文件，具体可参考用户登录成功发生界面跳转时动画的位置平移方法（本项目模块一的任务 3）。

模块三　新闻栏目列表功能的实现

【模块描述】

本模块为新闻栏目子项目，主要基于自定义新闻数据适配器、LinearLayout、ListView 组件、网络资源访问、图片异步加载、JSON 数据封装技术实现新闻栏目列表功能。

本模块将实现如下操作。

1）准备工作：测试服务器端新闻栏目列表 API 接口；
2）引用 title.xml 布局文件构建新闻栏目列表界面；
3）定义图片异步加载工具类；
4）在工具类 ParseJson 中定义解析新闻栏目列表数据的方法；

5) 创建新闻栏目列表的数据适配器 NewsListByNewsBarAdapter；
6) 实现新闻列表功能。

知识点	技能点
➢ 基于数据适配器填充 ListView 的方法 ➢ 功能模块化及代码重用设计思想 ➢ 自定义数据适配器 ➢ ListView 组件中列表样式的灵活定义 ➢ Android 基本 UI 组件运用 ➢ ListView 组件的 OnItemClickListener 的事件处理机制 ➢ 图片异步加载 ➢ Android 网络资源访问 ➢ JSON 轻量级数据封装	➢ 新闻栏目列表界面的实现 ➢ 工具类 ParseJson 中新闻栏目列表数据的 JSON 解析方法定义 ➢ 数据适配器 NewsListByNewsBarAdapter 的定义及 ListView 组件数据填充 ➢ 图片异步加载工具类的定义 ➢ 新闻栏目列表功能的实现

任务1 准备工作：测试服务器端新闻栏目列表 API 访问接口

【任务描述】

新闻客户端 APP 应用中新闻栏目列表的实现需要访问服务器端应用程序相关 API 接口，因此在实现新闻栏目列表功能之前，首先要确认服务器端应用程序是否部署成功并检验 API 接口是否可用。

【任务实施】

1) 按照模块 1 中任务 1 的方法确认服务器端 Tomcat 6.0 应用服务器成功启动。
2) 服务器端新闻栏目列表 API 接口带有新闻栏目列表参数，为如下地址：http://192.168.1.106:8080/News_PC/servlet/GetNewsByNewsBarAppServlet?newsBarID=1

其中"?"后面的"newsBarID"是需要传递的新闻栏目 id，"1"是新闻栏目 id 对应的参数值，意思是获取新闻栏目 id 为 1 的新闻列表。在浏览器地址栏中输入上述地址后按〈Enter〉键，出现如图 9-18 所示的界面，则 API 接口可用。

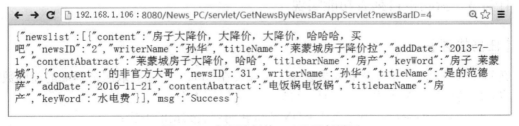

图 9-18 新闻访问接口 API 运行界面

上图中的数据来自服务器中的数据库，数据格式即为 JSON 数据格式。可以看出，上述所有新闻数据被封装在名称为"newslist"的 JSON 对象中，这个 JSON 对象是数组格式，里面封装了 2 个新闻栏目的条目，每个新闻栏目的条目包含 8 对键/值对，如第一个条目中 key 键 "writerName" 对应的 value 值为 "孙华"。在 APP 应用的新闻栏目列表展示功能中用编程实现对上述 JSON 数据进行解析的过程详见任务 3，具体实现参见任务 3。

任务 2　构建新闻栏目列表界面

【任务描述】

新闻栏目列表界面是将新闻栏目以列表的形式展现的界面。界面基于布局文件 title.xml 实现，代码更加简洁。另外，为了美化及定义新闻栏目列表项的外观和布局，需要创建一个列表项文件。本任务可以拆分为两个操作：

1) 创建列表项文件 news_list_item.xml；
2) 在新闻栏目列表界面中引用 title.xml，创建 activity_news_by_news_bar.xml 新闻栏目列表布局文件。

【任务实施】

1) 在 res 文件夹的 "layout" 中创建新闻栏目列表项布局文件 news_list_item.xml，界面布局采用两个线性管理器实现，嵌入一个 ImageView 组件以用来显示新闻图片，两个 TextView 组件用来显示新闻标题和发布时间，页面代码布局如图 9-19 所示。

2) 创建名称为 "NewsByNewsBarActivity" 的 Activity 类，"activity_news_by_news_bar.xml" 是界面布局文件，代码布局如图 9-20 所示。

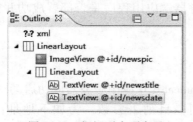
图 9-19　新闻列表项布局

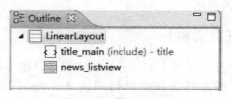
图 9-20　新闻列表界面布局文件

在线性布局管理器中有 include 和 ListView 两个组件，使用 include 可引用 title.xml 文件，即 < include layout = "@ layout/**title**" ... > 。

任务 3　新闻栏目列表功能实现

【任务描述】

新闻栏目列表功能的实现相对较复杂，不仅要在 APP 客户端获取服务器端的新闻数据，还要将获取的数据在列表中显示。具体可以细分为以下 4 个操作：

1) 在工具类 ParseJson 中定义从 JSON 解析新闻列表数据的方法；
2) 定义图片异步加载工具类 ImageTask；
3) 创建新闻列表数据适配器 NewsListByNewsBarAdapter；
4) 实现新闻列表功能。

【任务实施】

1）在"czmec.cn.shl.news.entity"包下创建 NewContent 类，用来封装新闻列表数据。

2）打开"czmec.cn.shl.news.engine"包下的工具类 ParseJson，定义方法为 getNewsList，解析从服务器获取的 JSON 新闻数据，其相应程序代码如下：

```
1.   //解析新闻列表
2.          public static List<NewsContent> getNewsList(String str){
3.             List<NewsContent> list = new ArrayList<NewsContent>();
4.             try{
5.                JSONObject jsonobject = new JSONObject(str);
6.                JSONArray dislist = jsonobject.getJSONArray("newslist");
7.                NewsContent news;
8.                for(int i=0;i<dislist.length();i++){
9.                   JSONObject jo = (JSONObject)dislist.opt(i);
10.                  news = new NewsContent();
11.                  news.setNewsID(jo.getString("newsID"));
12.                  try{
13.                     System.out.println("jo::::::::;" + jo.getString("newsID"));
14.                     news.setTitleName(jo.getString("titleName"));
15.                     news.setAddDate(jo.getString("addDate"));
16.                     news.setPic(jo.getString("pic"));
17.                  }catch(Exception e){
18.                     // TODO Auto-generated catch block
19.                     e.printStackTrace();
20.                  }
21.                  list.add(news);
22.               }
23.            }catch(JSONException e){
24.               e.printStackTrace();
25.               return null;
26.            }
27.            return list;
28.         }
```

3）定义图片异步加载工具类 ImageTask，其相应程序代码如下：

```
1.   public class ImageTask extends AsyncTask<Object,Object,Bitmap>{
2.      ImageView imageView = null;
3.      @Override
4.      protected Bitmap doInBackground(Object... params){
5.         //TODO Auto-generated method stub
6.         Bitmap bmp = null;
7.         imageView = (ImageView)params[1];
```

```
8.          try {
9.              bmp = BitmapFactory.decodeStream(new URL((String) params[0])
10.                     .openStream());
11.         } catch (Exception e) {
12.             //TODO Auto-generated catch block
13.             e.printStackTrace();
14.         }
15.         return bmp;
16.     }
17.
18.     protected void onPostExecute(Bitmap result) {
19.         imageView.setImageBitmap(result);
20.     }
21. }
```

上述程序代码中工具类 ImageTask 主要实现图片的异步加载功能,它继承了父类 AsyncTask<Object,Object,Bitmap>,重写了父类中的 Bitmap doInBackground(Object... params)和 onPostExecute(Bitmap result)方法,具体可参见本模块后面的"知识学习——AsyncTask 类"。

4)在"czmec.cn.shl.news.adapter"包下创建类 NewsListByNewsBarAdapter 新闻栏目列表数据适配器,其相应程序代码如下:

```
1.  public class NewsListByNewsBarAdapter extends BaseAdapter {
2.      private List<NewsContent> newsList;
3.      private LayoutInflater inflater;
4.      public NewsListByNewsBarAdapter(Context context, List<NewsContent> newsList)
5.      {
6.          inflater = LayoutInflater.from(context);
7.          this.newsList = newsList;
8.      }
9.      public int getCount() {
10.         //TODO Auto-generated method stub
11.         return newsList.size();
12.     }
13.     public Object getItem(int position) {
14.         //TODO Auto-generated method stub
15.         return position;
16.     }
17.     public long getItemId(int position) {
18.         return position;
19.     }
20.     public View getView(int position, View convertView, ViewGroup parent) {
21.         if (convertView == null)
```

```
22.        {
23.            convertView = inflater.inflate(R.layout.news_list_item,null);
24.        }
25.        TextView tvnewstitle = (TextView) convertView.findViewById(R.id.newstitle);
26.        tvnewstitle.setText(newsList.get(position).getTitleName());
27.        TextView tvnewsdate = (TextView) convertView.findViewById(R.id.newsdate);
28.        tvnewsdate.setText(newsList.get(position).getAddDate());
29.        ImageView c_img = (ImageView) convertView.findViewById(R.id.newspic);
30.        String pic = newsList.get(position).getPic();
31.        if(pic.equals(""))
32.        {
33.            c_img.setImageResource(R.drawable.ic_launcher);
34.        }
35.        else
36.        {
37.            new ImageTask().execute(Config.ImageAddress + pic,c_img);//异步加载
38.        }
39.        return convertView;
40.    }
41. }
```

上述程序代码第37行创建了图片异步加载类ImageTask，调用了execute方法实现图片异步加载。

5）打开"NewsByNewsBarActivity.java"类，修改新闻栏目列表项事件监听器setOnItemClickListener中界面跳转的代码：

```
Intent intent = new Intent(AllNewsBarActivity.this,AllNewsBarActivity.class);
```

6）打开"NewsByNewsBarActivity.java"类，实现新闻栏目列表功能，其相应程序代码如下：

```
1.  public class NewsByNewsBarActivity extends Activity {
2.      private ListView listview;
3.      private List<NewsContent> newsList;
4.      private NewsListByNewsBarAdapter adapter;
5.      private String newsBarID,newsBarName;
6.      protected void onCreate(Bundle savedInstanceState) {
7.          super.onCreate(savedInstanceState);
8.          setContentView(R.layout.activity_news_by_news_bar);
9.          Intent intent = getIntent();
10.         newsBarID = intent.getStringExtra("newsBarID");
11.         newsBarName = intent.getStringExtra("NewsBarName");
12.         listview = (ListView) this.findViewById(R.id.news_listview);
```

```
13.            TextView tv = (TextView) this.findViewById(R.id.title_name);
14.            tv.setText(newsBarName);
15.            new Thread(new Runnable() {
16.                public void run() {
17.                    getNewsByNewsBar();
18.                }
19.            }).start();
20.        }
21.        public void back(View v)
22.        {
23.            Intent intent = new Intent(NewsByNewsBarActivity.this, AllNewsBarActivity.class);
24.            startActivity(intent);
25.        }
26.        public void getNewsByNewsBar()
27.        {
28.            String url = Config.ServerAddress + "GetNewsByNewsBarAppServlet? newsBarID = " +
29. newsBarID;
30.            byte[] data;
31.            try {
32.                data = SynsHttp.readParse(url);
33.                String jsonStr = new String(data);
34.                newsList = ParseJson.getNewsList(jsonStr);
35.                for(NewsContent newsbar : newsList)
36.                {
37.                    System.out.println("打印:" + newsbar.getTitleName());
38.                }
39.            } catch (Exception e) {
40.                e.printStackTrace();
41.            }
42.            runOnUiThread(new Runnable() {
43.                public void run() {
43.                    adapter = new NewsListByNewsBarAdapter(NewsByNewsBarActivity.this,
44. newsList);
45.                    listview.setAdapter(adapter);
46.                    listview.setOnItemClickListener(new OnItemClickListener() {
47.                        public void onItemClick(AdapterView<?> arg0, View arg1, int arg2,
48.                            long arg3) {
49.                            System.out.println(" ----- 被选中的 Item 是 ---->>" + arg2);
50.                            Toast.makeText(NewsByNewsBarActivity.this, "你选中的新闻是:
51. id:" + newsList.get(arg2).getNewsID() + ",新闻名称是为" + newsList.get(arg2).getTitleName(),
52. 4000).show();
53.                        }
54.                    });
55.                }
```

56.　　　　});
57.　　　}
58.　}

当选择图 9-17 所示的新闻栏目列表界面中的"军事"栏目时，将进入"军事"新闻列表界面，如图 9-21 所示。

【知识学习】AsyncTask 类

在 Android 中实现异步任务机制有两种方式，Handler 和 AsyncTask。Handler 模式需要为每一个任务创建一个新的线程，任务完成后通过 Handler 实例向 UI 线程发送消息，完成界面的更新，这种方式对整个过程的控制比较精细。但也是有缺点的，例如代码相对臃肿，在多个任务同时执行时，不易对线程进行精确的控制。为了简化操作，Android 1.5 提供了工具类 android.os.AsyncTask，它使创建异步任务变得更加简单，不再需要编写任务线程和 Handler 实例即可完成相同的任务。

图 9-21　新闻列表运行界面

一个异步任务的执行一般包括以下几个步骤。

1）execute(Params...params)：执行一个异步任务，需要在代码中调用此方法，触发异步任务的执行。

2）onPreExecute：在 execute(Params...params) 被调用后立即执行，一般在执行后台任务前对 UI 做一些标记。

3）doInBackground(Params...params)：在 onPreExecute 完成后立即执行，用于执行较为费时的操作，此方法将接收输入参数和返回计算结果。在执行过程中可以调用 publishProgress(Progress...values) 来更新进度信息。

4）onProgressUpdate(Progress...values)：在调用 publishProgress(Progress...values) 时，此方法被执行，直接将进度信息更新到 UI 组件上。

5）onPostExecute(Result result)：当后台操作结束时，此方法将会被调用，计算结果将作为参数传递到此方法中，直接将结果显示到 UI 组件上。

【任务拓展】

当选择图 9-21 所示的军事新闻栏目列表中的列表选项时，界面跳转到对应的军事新闻详细信息界面。

提示：

1）创建新闻栏目详细界面 NewsDetailActivity；

2）在军事栏目列表界面中的列表项事件监听器中获取被选中的新闻 id，将当前界面跳转到新闻栏目详细界面，并传递新闻 id 参数；

3）在新闻栏目详细界面 NewsDetailActivity 获取传递过来的新闻 id 参数，根据 id 获取新闻；

4）将新闻栏目列表相关信息显示在 NewsDetailActivity 界面。

参 考 文 献

[1] 李刚. 疯狂 Android 讲义［M］. 2 版. 北京：电子工业出版社，2013.
[2] 明日科技. Android 从入门到精通［M］. 北京：清华大学出版社，2012.
[3] Android 编程权威指南［M］. 2 版. 王明发，译. 北京：人民邮电出版社，2016.
[4] 陈文. 深入理解 Android 网络编程［M］. 北京：机械工业出版社，2013.
[5] Reto Meier. Android4 高级编程［M］. 3 版. 余建伟，赵凯，译. 北京：清华大学出版社，2013.